T0339918

Near-infrared Speckle Contrast Diffuse Correlation Tomography for Noncontact Imaging of Tissue Blood Flow Distribution

Imaging of tissue blood flow (BF) distributions provides vital information for the diagnosis and therapeutic monitoring of various vascular diseases. The innovative near-infrared speckle contrast diffuse correlation tomography (scDCT) technique produces full 3D BF distributions. Many advanced features are provided over competing technologies including high sampling density, fast data acquisition, noninvasiveness, noncontact, affordability, portability, and translatability across varied subject sizes. The basic principle, instrumentation, and data analysis algorithms are presented in detail. The extensive applications are summarized such as imaging of cerebral BF (CBF) in mice, rat, and piglet animals with skull penetration into the deep brain. Clinical human testing results are described by recovery of BF distributions on preterm infants (CBF) through incubator walls, and on sensitive burn tissues and mastectomy skin flaps without direct device–tissue interactions. Supporting activities outlined include integrated capability for acquiring surface curvature information, rapid 2D BF mapping, and optimizations via tissue-like phantoms and computer simulations. These applications and activities both highlight and guide the reader as to the expected abilities and limitations of scDCT for adapting into their own preclinical/clinical research, use in constrained environments (i.e., neonatal intensive care unit bedside), and use on vulnerable subjects and measurement sites.

Near-infrared Speckle Contrast Diffuse Correlation Tomography for Noncontact Imaging of Tissue Blood Flow Distribution

Daniel Irwin, Siavash Mazdeyasna,
Chong Huang, Mehrana Mohtasebi, Xuhui Liu,
Lei Chen and Guoqiang Yu

CRC Press
Taylor & Francis Group
Boca Raton London New York

CRC Press is an imprint of the
Taylor & Francis Group, an **informa** business

First edition published 2022
by CRC Press
6000 Broken Sound Parkway NW, Suite 300, Boca Raton, FL 33487-2742

and by CRC Press
4 Park Square, Milton Park, Abingdon, Oxon, OX14 4RN

CRC Press is an imprint of Taylor & Francis Group, LLC

ISBN: 9781032133874 (hbk)
ISBN: 9781032159362 (pbk)
ISBN: 9781003246374 (ebk)

DOI: 10.1201/9781003246374

Typeset in Times
by codeMantra

Contents

Acknowledgments or Credits List

We would like to thank the following collaborators for their contributions toward the research, development, and experimentation of speckle contrast diffuse correlation tomography (scDCT): Yu Lin, PhD; Yu Shang, PhD; Lian He, PhD; Ahmed A. Bahrani, PhD; Mingjun Zhao, PhD; Weikai Kong, MS; Ting Li, PhD; Yutong Gu, MEng; Yanda Cheng; Jing Chen, PhD; Myeongsu Seong, PhD; Joshua Paul Morgan, MEng; Jae Gwan Kim, PhD; Jia Luo, PhD; Lesley Wong, MD; Nneamaka B. Agochukwu, MD; Li Chen, PhD; Elie G. Abu Jawdeh, MD; Henrietta S. Bada, MD; Kathryn E. Saatman, PhD; Alisha Bonaroti, MD; Ryan C. DeCoster, MD; Qiang Cheng, PhD; Patrick McGrath, MD; Jeffrey Todd Hastings, PhD; Margaret M. Szabunio, MD; Jeffrey P. Radabaugh, MD; Rony K. Aouad, MD; Thomas J. Gal, MD; Amit B. Patel, MD; and Joseph Valentino, MD.

The studies in this book were made possible with assistance from the following funding sources: National Institutes of Health (NIH, R01-EB028792, R01-HD101508, R21-HD091118, R21NS114771, R41-NS122722, and R56-NS117587), American Heart Association (AHA, #16GRNT30820006, #14SDG20480186, and Predoctoral Fellowship #835726), and National Science Foundation (NSF, #1539068). The content is solely the responsibility of the authors and does not necessarily represent the official views of the NIH, AHA, or NSF.

There have been two patents issued for scDCT-related technologies which include: US Patent #9861319 and US Patent #10842422.

Authors

Daniel Irwin is a Research Scientist in the Department of Biomedical Engineering at the University of Kentucky. He has been contributing to the Biomedical Optics field for nine years, completing his PhD in 2018 with dissertation work including efforts toward scDCT development, analysis, and imaging. His current research involves advancing near-infrared diffuse speckle-based spectroscopic and imaging instrumentation.

Siavash Mazdeyasna received his PhD in Biomedical Engineering from the University of Kentucky (UK) in the fall of 2020. His research during his doctoral and postdoctoral studies at the UK has provided him with extensive experience in biomedical engineering especially in developing the state-of-the-art diffuse optical imaging system. He has worked on both hardware and software development, data analysis, and validations in tissue-simulating phantoms, small animals, and human subjects.

Chong Huang worked as a Postdoctoral Scholar and then Research Assistant Professor in the Department of Biomedical Engineering at the University of Kentucky. Over the past nine years, he worked on the development of various near-infrared diffuse optical spectroscopic and tomographic technologies for noninvasive assessment of tissue hemodynamics and metabolism. He is the co-inventor of the scDCT technology.

Mehrana Mohtasebi is a PhD candidate and American Heart Association Predoctoral Fellow in the Department of Biomedical Engineering at the University of Kentucky. Mehrana has been working on optimization and validation of scDCT against established techniques for quantifying resting-state functional connectivity networks in animal models of stroke and intraventricular hemorrhage to provide objective assessments of brain tissue injury.

Xuhui Liu is a PhD candidate in the Department of Biomedical Engineering at the University of Kentucky. He obtained his bachelor's degree in Electrical Engineering at the University of Electronic Science and Technology of China. His research focuses on developing a wearable fiber-free optical sensor to continuously monitor cerebral blood flow and oxygenation in different subjects (mice, piglets, preterm infants).

Lei Chen is an Associate Professor in the Department of Physiology, Spinal Cord and Brain Injury Research Center, and the Department of Biomedical Engineering at the University of Kentucky. His research focuses on the activation of stem cells and other reparative mechanisms after a head injury, including ischemic stroke, head trauma, and neonatal intraventricular hemorrhage. Those animal models have been utilized for the validation of scDCT to achieve the continuous and longitudinal monitoring of cerebral injury and recovery.

Guoqiang Yu is a Professor in the Department of Biomedical Engineering at the University of Kentucky. Over the past decade, he leads the Biomedical Optics Laboratory where his group has invented several highly powerful functional imaging technologies, including scDCT for noncontact imaging of brains, burns, wounds, and reconstructive tissues, a wearable fiber-free optical sensor for continuous cerebral monitoring in freely behaving subjects, wearable fluorescence eye loupes, and a time-resolved laser speckle contrast imaging system. These patented technologies and devices hold great promise for imaging numerous organs in preclinical and clinical settings. He has established a track record of publications in high-impact journals. Particularly, he has written a book chapter in the *Handbook of Biomedical Optics* (2011, CRC Press), titled "Near-infrared diffuse correlation spectroscopy (DCS) for assessment of tissue blood flow".

Introduction

1

SIGNIFICANCE OF BLOOD FLOW IMAGING

Tissue blood flow (BF) provides vital information for the detection, diagnosis, therapeutic monitoring, treatment, and study of vascular-impacted diseases. BF status and change can be contextually interpreted for many purposes. Contributors to observed variations include subject health, disease presence and progression, physical activity and performance, and vessel genesis, modification, constriction, and destruction. These environments affect the delivery of oxygen, nutrients, drugs, and other endogenous (i.e., oxygenated and deoxygenated hemoglobin) and exogenous (i.e., indocyanine green (ICG)) blood-borne agents. Modern BF imaging (BFI) instruments have naturally manifested in alignment with these physiological considerations. They are inherently framed inside available technologies while offering their own unique utility, hitting reachable healthcare issues, and resolving competing detrimental factors. To understand BFI devices in general and within the perspective of our primary system, speckle contrast diffuse correlation tomography (scDCT), we first summarize BF as a biomarker.

BF study parallels that carried out for other biomarkers. Healthy tissues of proper composition and function frequently serve as a baseline for identifying deviations from an expected, normal state. Observed effects are often dichotomized into temporal and spatial aspects. Time course evolutions of muscle BF distributions can express localized responses to occlusion (e.g., cuff), tasks (e.g., exercise), revascularization procedures (e.g., peripheral arterial disease), and physical influencers (e.g., temperature and pressure). Abnormal spatial BF distributions can occur due to tumor growth, physically or chemically injured tissue sites (i.e., burns and wounds), surgical procedures (i.e., ligations, mastectomy), or compromised brain autoregulation. Access to, and control of, both aspects reveal a more complete view of the underlying processes. As our interest is in BFI, we are directed toward all applications wherein spatiotemporal monitoring of abnormal or changing BF harnesses significant information directly or indirectly. It is because of BFs' widely splayed grasp, both in a semi-literal physiological as well as medically useful sense, that the continuous advancement of BFI devices is so critical.

TECHNOLOGIES FOR BLOOD FLOW IMAGING

Devices engineered for research and clinical BFI operation are comparable along numerous facets. Probing sources (e.g., ultrasound waves, x-rays, radio waves, and optics) dictate anticipated probe-tissue interactions that result in source signal attenuation, dominant material/tissue sensitivity, and safety concerns (e.g., ionization). Detection mechanisms insert another layer in the signal path further affecting attenuation and signal sensitivity. Different combinations of these fundamental components produce other characteristics of the resulting systems. Among these include instrumentation cost, operability, size, potential subject groups, translatability, portability, and resolution (temporal and spatial). With these modern BFI systems' polychotomous features in hand, we can now outline how scDCT arranges itself within.

Near-infrared (NIR) light serves as the probing source for scDCT. NIR medical imaging systems fall under the umbrella term of biomedical optics. Developments over the past century, especially the last few decades, have promoted ever-increasing interest in NIR devices. Injecting beams of non-ionizing photons is a safe alternative to some of the traditional probing sources. These photons, with the help of the so-called NIR window, have been found to penetrate tissues up to several centimeters deep. Many soft tissues (e.g., muscle and brain) also influence photons to scatter many times and propagate diffusely. That is, introducing an NIR light source at the tissue surface and placing a detection element nearby facilitates noninvasive reflectance measurements. If a sample is sufficiently thin, transmission setups may also be feasible. For scDCT, we are specifically engaged in the NIR diffusive regime of biological tissues.

Diffuse NIR optical techniques are broadly used to recover deep tissue optical/hemodynamic properties (up to several cm). Near-infrared spectroscopy (NIRS) is used to obtain tissue oxygenation measurements and diffuse correlation spectroscopy (DCS) to obtain tissue BF [1–6]. NIRS/DCS technologies have been extensively reviewed in many publications, including the *Handbook of Biomedical Optics* [3,7–11]. NIRS/DCS systems utilize a limited number of discrete sources and detectors and thus lack the combination of spatiotemporal resolution and wide field-of-view (FOV) to image spatially distributed tissue functions.

There have been recent advancements toward using point illumination for deep tissue penetration and CCD/CMOS cameras for high-density 2D detection of spatial fluctuations of diffuse laser speckle contrast to facilitate rapid high-density imaging of BF distributions in deep tissues. From 2013 to 2014, two techniques emerged relating quantification of diffuse speckle contrast with traditional DCS theory. These spectroscopic implementations included diffuse speckle contrast analysis (DSCA) and speckle contrast optical spectroscopy (SCOS) [12–14]. In early 2014, the transmission-based imaging technique of speckle contrast optical tomography (SCOT) used a direct analytical relationship between the mean-square-displacement (MSD) of moving scatterers and diffuse speckle contrasts in tissue-simulating phantoms [15]. By 2015, our reflectance- and finite element method (FEM)-based scDCT technique was created opening a window for studies in human subjects where deep/large tissue volumes can be imaged [16]. Together, these advanced CCD/CMOS-based techniques (e.g., DSCA, SCOS, SCOT, and scDCT) are inherently founded upon the same concept despite differences in nomenclature and technological evolution. They have subsequently been applied and improved upon in many ways since their origination, including reported imaging of BF distributions in animal models and humans [12,13,15–31].

Our patented scDCT (US Patent #9861319, 2016) is a significant step above our original partial contact model from 2015. A galvo-mirror remotely delivers coherent point NIR light to source positions and a CCD/CMOS camera measures spatial diffuse speckle contrasts on the tissue boundary. These boundary data are simply averaged for 2D mapping of surface BF or input into our patented FEM-based program for 3D reconstruction of BF distributions in deep tissue volumes [16,23–31]. The innovative scDCT provides many unique advanced features over other competitive technologies, including noncontact imaging technique; high sampling density; fast data acquisition; adjustable FOV for translational studies; and an affordable, portable device. The scDCT has been extensively tested for imaging of BF distributions in animal brains (mice, rats, and piglets) and human tissues (infant brains, wound/burned tissues, and mastectomy skin flaps) [16,23–32].

There are also optically relevant parameters that arise throughout the text due to their usefulness and/or inclusion under theoretical topics. These are tissue optical properties (absorption, μ_a, scattering, μ_s') and arterial blood oxygen saturation (SaO$_2$).

HISTORY OF SCDCT DEVELOPMENT

The history of scDCT is visualized from both instrumental and application perspectives in Figure 1.1. We work forward instrumentally starting from our portable dual-wavelength DCS flow-oximeter created in 2009 [3]. This device pulled together NIRS and DCS techniques for simultaneously acquiring relative chromophore concentration (oxygenated hemoglobin, $\Delta[HbO_2]$, deoxygenated hemoglobin $\Delta[Hb]$) and relative BF (rBF; see Chapter 2) changes. This hybrid instrument provided an inexpensive tissue hemodynamic monitoring option with sufficient portability for bedside usage. Our noncontact dual-wavelength DCS (ncDCS) system followed in 2012, recovering oxygenation and BF changes without requiring a probe directly on the tissue surface [33,34]. In ncDCS, an optical lens system projected light sources onto tissue, collected emitted light onto a detector, and provided flexible selection of SD separation. Independent source/detector optical paths reduced interference of source light reflections on detected light. This system opened measurement applications to tissues that may be sensitive (e.g., ulcers) or easily compressible (e.g., breast) while still benefiting from safe NIR light. Compressed tissues can result in their altered optical properties and BF. At the time, both the portable flow-oximeter and ncDCS were still limited to a small set of spatial data points due to limited numbers of discrete sources and detectors.

Efforts continued in the direction of creating an instrument capable of recovering data from a large region of interest (ROI) and implementing tomographic 3D imaging. This was accomplished by 2014 with the noncontact diffuse correlation tomography (ncDCT) system, focused on flow imaging [35]. In ncDCT, the lens focusing apparatus from ncDCS was significantly improved upon to house a linear array of detectors (e.g., detector fiber tips) with independent source paths (e.g., source fiber tips) at each end. This apparatus was affixed to a motorized linear (or angular) mechanical stager that could be focused onto the tissue surface and automatically scan over an ROI at chosen step sizes. A new flow reconstruction method, leveraging existing techniques,

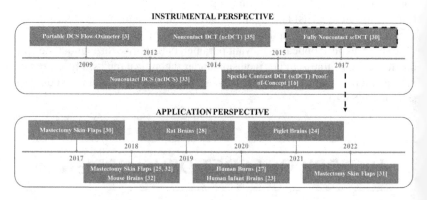

FIGURE 1.1 Historical timeline of scDCT from instrumental and application perspectives.

enabled rapid integration into an open-source reconstruction software package (NIRFAST) [36]. This first-generation BFI device offered a flexible, portable solution for tomographic (or spectroscopic) applications such as mastectomy skin flaps, head and neck skin flaps, and breast tumors [37–39]. The higher spatial resolution and tomographic imaging supported greater insight into regional physiological BF activities. Limitations of this setup included complexity in aligning the source-detector arrangements on tissues [40], lack of scaling the ROI size with subject or measurement area size, freedom of spatial density restricted to one dimension (direction of linear stager; arc for rotational stager), and long scan times for each data frame.

scDCT evolved to remedy many of the aforementioned constraints with our first partial contact, proof-of-concept (PoC) system completed in 2015 [16]. This unit replaced the static detector fiber array of ncDCT with a highly sensitive electron-multiplying charge-coupled device (EMCCD) camera. A two-dimensional (2D) grid of pixels acted as a set of densely grouped detection elements (over 1 million). Subsets were selected as representative logical detectors based on experimental usage. The quantitation of flow was possible through introduction of diffuse correlation theories into existing, but traditionally superficially sensitive, laser speckle contrast imaging (LSCI) theories [41,42]. Reconstruction methods as developed for ncDCT needed no additional effort outside of boundary data conversion. An optical switch sequenced laser light among four source fibers circling the exposure area in a restrictive, contact manner. A full noncontact version was subsequently created by 2017, replacing the switch with a 2D scanning galvo-mirror and accompanied polarizers as highlighted in the previous section [30]. In this text when we refer to scDCT, it is in reference to this last fully noncontact base instrumentation unless otherwise noted.

In the next chapter, we examine how to use scDCT for BFI in greater technical detail. Chapter 3 lays out the applications of scDCT, depicted by the timeline of Figure 1.1, in tissue-like phantoms, by computer simulations, and in multiple human and animal tissues.

REFERENCES

1. Shang, Y., et al., Cerebral monitoring during carotid endarterectomy using near-infrared diffuse optical spectroscopies and electroencephalogram. Phys Med Biol, 2011. **56**: p. 3015–3032.
2. Cheng, R., et al., Noninvasive optical evaluation of spontaneous low frequency oscillations in cerebral hemodynamics. Neuroimage, 2012. **62**(3): p. 1445–1454.

3. Shang, Y., et al., Portable optical tissue flow oximeter based on diffuse correlation spectroscopy. Opt Lett, 2009. **34**(22): p. 3556–3558.
4. Gurley, K., Y. Shang, and G. Yu, Noninvasive optical quantification of absolute blood flow, blood oxygenation, and oxygen consumption rate in exercising skeletal muscle. J Biomed Opt, 2012. **17**(7): p. 075010.
5. Shang, Y., et al., Extraction of diffuse correlation spectroscopy flow index by integration of *N*th-order linear model with Monte Carlo simulation. Appl Phys Lett, 2014. **104**(19): p. 193703.
6. Cheng, R., et al., Near-infrared diffuse optical monitoring of cerebral blood flow and oxygenation for the prediction of vasovagal syncope. J Biomed Opt, 2014. **19**(1): p. 17001.
7. Yu, G., et al., Near-infrared diffuse correlation spectroscopy (DCS) for assessment of tissue blood flow, in Handbook of Biomedical Optics, D. Boas, C. Pitris, and N. Ramanujam, Editors. 2011, Boca Raton, FL: Taylor & Francis Books Inc. p. 195–216.
8. Yu, G., Diffuse Correlation Spectroscopy (DCS): A diagnostic tool for assessing tissue blood flow in vascular-related diseases and therapies. Curr Med Imaging Rev, 2012. **8**(3): p. 194–210.
9. Buckley, E.M., et al., Diffuse correlation spectroscopy for measurement of cerebral blood flow: Future prospects. Neurophotonics, 2014. **1**(1): p. 011009.
10. Boushel, R., et al., Monitoring tissue oxygen availability with near infrared spectroscopy (NIRS) in health and disease. Scand J Med Sci Sports, 2001. **11**(4): p. 213–222.
11. Scheeren, T.W., P. Schober, and L.A. Schwarte, Monitoring tissue oxygenation by near infrared spectroscopy (NIRS): Background and current applications. J Clin Monit Comput, 2012. **26**(4): p. 279–287.
12. Bi, R., J. Dong, and K. Lee, Deep tissue flowmetry based on diffuse speckle contrast analysis. Opt Lett, 2013. **38**(9): p. 1401–1403.
13. Bi, R., J. Dong, and K. Lee, Multi-channel deep tissue flowmetry based on temporal diffuse speckle contrast analysis. Opt Express, 2013. **21**(19): p.22854–22861.
14. Valdes, C.P., et al., Speckle contrast optical spectroscopy, a non-invasive, diffuse optical method for measuring microvascular blood flow in tissue. Biomed Opt Express, 2014. **5**(8): p. 2769–2784.
15. Varma, H.M., et al., Speckle contrast optical tomography: A new method for deep tissue three-dimensional tomography of blood flow. Biomed Opt Express, 2014. **5**(4): p. 1275–1289.
16. Huang, C., et al., Speckle contrast diffuse correlation tomography of complex turbid medium flow. Med Phys, 2015. **42**(7): p. 4000–6.
17. Bi, R., et al., Optical methods for blood perfusion measurement--theoretical comparison among four different modalities. J Opt Soc Am A Opt Image Sci Vis, 2015. **32**(5): p. 860–866.
18. Seong, M., et al., Simultaneous blood flow and blood oxygenation measurements using a combination of diffuse speckle contrast analysis and near-infrared spectroscopy. J Biomed Opt, 2016. **21**(2): p. 27001.
19. Lee, K., Diffuse Speckle Contrast Analysis (DSCA) for Deep Tissue Blood Flow Monitoring. Adv Biomed Eng, 2020. **9**: p. 21–30.

20. Yeo, C., et al., Low frequency oscillations assessed by diffuse speckle contrast analysis for foot angiosome concept. Sci Rep, 2020. **10**(1): p. 17153.

21. Dragojevic, T., et al., High-density speckle contrast optical tomography (SCOT) for three dimensional tomographic imaging of the small animal brain. Neuroimage, 2017. **153**: p. 283–292.

22. Dragojevic, T., et al., High-density speckle contrast optical tomography of cerebral blood flow response to functional stimuli in the rodent brain. Neurophotonics, 2019. **6**(4): p. 045001.

23. Abu Jawdeh, E.G., et al., Noncontact optical imaging of brain hemodynamics in preterm infants: A preliminary study. Phys Med Biol, 2020. **65**(24): p. 245009.

24. Huang, C., et al., Speckle contrast diffuse correlation tomography of cerebral blood flow in perinatal disease model of neonatal piglets. J Biophotonics, 2021. **14**(4): p. e202000366.

25. Mazdeyasna, S., et al., Noncontact speckle contrast diffuse correlation tomography of blood flow distributions in tissues with arbitrary geometries. J Biomed Opt, 2018. **23**(9): p. 1–9.

26. Bonaroti, A., et al., The role of intraoperative laser speckle imaging in reducing postoperative complications in breast reconstruction. Plast Reconstr Surg, 2019. **144**(5): p. 933e–934e.

27. Zhao, M., et al., Noncontact speckle contrast diffuse correlation tomography of blood flow distributions in burn wounds: A preliminary study. Mil Med, 2020. **185**(Suppl 1): p. 82–87.

28. Huang, C., et al., Noninvasive noncontact speckle contrast diffuse correlation tomography of cerebral blood flow in rats. Neuroimage, 2019. **198**: p. 160–169.

29. Yu, G., Y. Lin, and C. Huang, Noncontact three-dimensional diffuse optical imaging of deep tissue blood flow distribution. U.S. Patent 9,861,319.

30. Huang, C., et al., Noncontact 3-D speckle contrast diffuse correlation tomography of tissue blood flow distribution. IEEE Trans Med Imaging, 2017. **36**(10): p. 2068–2076.

31. Mazdeyasna, S., et al., Intraoperative optical and fluorescence imaging of blood flow distributions in mastectomy skin flaps for identifying ischemic tissues. Plast Reconstr Surg, 2022. **150**(2): p. 282–287.

32. Mazdeyasna, S., et al., Noninvasive noncontact 3D optical imaging of blood flow distributions in animals and humans. IEEE International Symposium on Signal Processing and Information Technology, 2018: p. 441–446.

33. Lin, Y., et al., Noncontact diffuse correlation spectroscopy for noninvasive deep tissue blood flow measurement. J Biomed Opt, 2012. **17**(1): p. 010502.

34. Li, T., et al., Simultaneous measurement of deep tissue blood flow and oxygenation using noncontact diffuse correlation spectroscopy flow-oximeter. Sci Rep, 2013. **3**: p. 1358.

35. Lin, Y., et al., Three-dimensional flow contrast imaging of deep tissue using noncontact diffuse correlation tomography. Appl Phys Lett, 2014. **104**(12): p. 121103.

36. Dehghani, H., et al., Near infrared optical tomography using NIRFAST: Algorithm for numerical model and image reconstruction. Commun Numer Methods Eng, 2008. **25**(6): p. 711–732.

37. Huang, C., et al., Noncontact diffuse optical assessment of blood flow changes in head and neck free tissue transfer flaps. J Biomed Opt, 2015. **20**(7): p. 075008.

38. He, L., et al., Noncontact diffuse correlation tomography of human breast tumor. J Biomed Opt, 2015. **20**(8): p. 86003.

39. Agochukwu, N.B., et al., A Novel noncontact diffuse correlation spectroscopy device for assessing blood flow in mastectomy skin flaps: A prospective study in patients undergoing prosthesis-based reconstruction. Plast Reconstr Surg, 2017. **140**(1): p. 26–31.

40. Huang, C., et al., Alignment of sources and detectors on breast surface for noncontact diffuse correlation tomography of breast tumors. Appl Opt, 2015. **54**(29): p. 8808–8816.

41. Fercher, A.F. and J.D. Briers, Flow visualization by means of single-exposure speckle photography. Opt Commun, 1981. **37**(5): p. 326–330.

42. Bandyopadhyay, R., et al., Speckle-visibility spectroscopy: A tool to study time-varying dynamics. Rev Sci Instrum, 2005. **76**(9): p. 093–110.

scDCT Methods

2

PRINCIPLE

In order for a medical device to interrogate a sample, generally three basic entities must exist: a present or added interrogating-capable probe, matter which interacts with the probe in transit, and a probe collector. For scDCT, the probe is coherent near-infrared (NIR) light, the material is biological tissue (or phantoms) up to several cm depth, and the collection is performed by optically sensitive transducers. Specific interactors within biological tissues are the dominant scatterers (e.g., organelles, collagen, and mitochondrion) and absorbers (e.g., hemoglobin, water, and lipids). Interactors within phantoms vary, which allows some flexibility in material selection but standardization is not available for every option.

There are several theories that together govern the scope and context of scDCT usage. The traditional modalities from which scDCT most directly extends are diffuse correlation tomography (DCT) and laser speckle contrast imaging (LSCI). Both follow dynamically changing speckle of coherent light for the purposes of quantifying BF, usually through an index (DCT) or contrast coefficient (LSCI). From these precursors, the connection between them is made to formulate support for our scDCT algorithm in acquiring internal BF data from boundary measurements.

DCS is a deep tissue BF investigative technique that underlies the DCT framework. DCS itself extends from the concept of dynamic light scattering (DLS) [1–4]. It is modeled theoretically by the correlation diffusion equation (CDE). The CDE (Equation 2.1) describes propagation of the temporal electrical field autocorrelation function, $G_1(r,\tau) = \langle E(r,t)E^*(r,t+\tau)\rangle$, in turbid samples over a correlation decay time τ (seconds) with position vector r (cm), source $S(r)$, speed of light in the medium v (cm/s), photon diffusion coefficient

DOI: 10.1201/9781003246374-2

$D(r) \approx v/3\mu_s'(r)$, reduced scattering coefficient μ_s' (1/cm), total absorption coefficient $\mu_a^{\text{total}}(r,\tau) = \mu_a(r) + \frac{1}{3}\alpha\mu_s'(r)k_0^2\langle\Delta r^2(r,\tau)\rangle$, absorption coefficient μ_a (1/cm), moving to total scatterer ratio α, wavenumber k_0, and mean-square-displacement (MSD) of scattering particles $\langle\Delta r^2(r,\tau)\rangle$ [4]. The primary moving scatterers in biological tissues are red blood cells (RBC) and in tissue-like phantoms are additives such as intralipid emulsions.

$$\left[\nabla \cdot D(r)\nabla - v\mu_a^{\text{total}}(r,\tau)\right]G_1(r,\tau) = -vS(r) \tag{2.1}$$

Studies in this text model MSD as Brownian motion with $\langle\Delta r^2(r,\tau)\rangle = 6\tau D_B$ where D_B is an effective diffusion coefficient of moving scatterers. Other MSD models may be found in the literature. As hardware described in the next section generates a collimated light beam, an approximate diffuse source for the CDE is created by moving the collimated source one scattering distance into a sample, $1/\mu_s'$ [5]. The measure of BF is identified by resolving the combined αD_B term. This term has been found to give a representative BF response as validated against power spectral Doppler ultrasound, Doppler ultrasound, arterial-spin-labeled MRI, fluorescent microspheres, Xenon CT, and laser Doppler flowmetry [6–13]. In most studies, rBF changes are reported in reference to some baseline, $(\alpha D_B)_0$, as $\text{rBF} = \alpha D_B/(\alpha D_B)_0$.

In samples shaped with a semi-infinite geometry, Equation 2.2 is the solution to the CDE with boundary source-detector (SD) separation from a source position r_s as $\rho = |r - r_s|$, isotropic source depth $z_0 = 1/\mu_s'$ and separation $r_1 = \sqrt{\rho^2 + z_0^2}$, negative isotropic imaging source depth $z_b = 2(1 + R_{\text{eff}})/3\mu_s'(1 - R_{\text{eff}})$ and separation $r_2 = \sqrt{\rho^2 + (z_0 + 2z_b)^2}$, refractive index mismatch $R_{\text{eff}} = -1.440n^{-2} + 0.710n^{-1} + 0.0636n + 0.668$, and $M^2 = 3\mu_a\mu_s' + \mu_s'^2k_0^2\alpha\langle\Delta r^2(\tau)\rangle$ [14]. We treat the MSD as before, $\langle\Delta r^2(\tau)\rangle = 6\tau D_B$, and assume $n \approx 1.33$ for tissues and tissue-like phantoms. From this solution, a particular αD_B is quantified with regards to a single SD distance.

$$G_1(\rho,\tau) = \frac{vS_0}{4\pi D}\left[\frac{\exp(-r_1 M)}{r_1} - \frac{\exp(-r_2 M)}{r_2}\right] \tag{2.2}$$

DCT and ncDCT imaging consist of acquiring many such SD measurements. This boundary data is fed into an image reconstruction algorithm, along with other relevant factors, to produce a tomographic BF distribution. The image reconstruction technique, if using boundary αD_B, for ncDCT and scDCT is equivalent. With the boundary data measurement methods for DCT/ncDCT, collecting dense SD data sets can be prohibitively expensive (e.g., avalanche photodiodes), scale poorly (e.g., optical fibers connected to each SD endpoint), and lack translatable efficiency. LSCI techniques as described next open the path for remedying these concerns.

LSCI is commonly used as a noninvasive, superficially sensitive NIR BF measurement technique supplied with a wide-field light source and multipixel camera sensor. Flow variations are monitored through a measure of speckle contrast, K_s, quantified according to Equation 2.3 with spatial (or temporal) standard deviation σ_s and expected value $\langle I \rangle$ of a set of intensities within some local region [15]. A generalization of this procedure is described by speckle-visibility-spectroscopy (SVS), quantifying the observability of dynamic speckle patterns [16]. As in the case of DCS/DCT, normalized BF to a baseline, $(K_s)_0$, is performed by rBF $= K_s/(K_s)_0$.

$$K_s(r) = \frac{\sigma_s}{\langle I \rangle} \tag{2.3}$$

The relation between speckle contrast, $K_s(r)$, and the normalized temporal electric field autocorrelation function, $g_1(r,\tau) = G_1(r,\tau)/G_1(r,0)$, is given by Equation 2.4 (assuming Siegert relation) with exposure time T and β as a factor for system optics [16].

$$K_s^2(r) = \frac{2\beta}{T} \int_0^T \left(1 - \frac{\tau}{T}\right) g_1^2(r,\tau)\, d\tau \tag{2.4}$$

The Nyquist spatial sampling criterion requires $\rho_{\text{speckle}} > 2\rho_{\text{pixel}}$ with speckle size ρ_{speckle} and pixel size ρ_{pixel}. Speckle size varies, but the lower limit spot size (used for Nyquist calculations) is related to the light source and system optics by Equation 2.5 with magnification M and f-number $f/\#$ [17,18].

$$\rho_{\text{speckle}} = 2.44\lambda(1+M)f/\# \tag{2.5}$$

For scDCT, the normalized form of g_1 from the semi-infinite CDE solution is used in the integral definition of K_s^2 to establish a nonlinear relationship (Equation 2.6) between the LSCI-based speckle contrast and DCT-based flow index [19]. In Equation 2.6, $C = \left[r_2 \exp(-r_1 a) - r_1 \exp(-r_2 a) \right]^{-1}$,

$$A = \frac{1}{3\mu_s'^2 k_0^2 \alpha D_B} \sum_{i=1}^{3} \left(\frac{x_i}{y_i^2}\right)\left[(y_i b - 1)\exp(y_i b) - (y_i a - 1)\exp(y_i a)\right],$$

$$B = \frac{1}{18\mu_s'^4 k_0^4 (\alpha D_B)^2} \sum_{i=1}^{3} \left(\frac{x_i}{y_i^4}\right)\left[\begin{array}{l}\left(y_i^3 b^3 - 3y_i^2 b^2 - y_i^3 b a^2 + 6y_i b + y_i^2 a^2 - 6\right),\\ \exp(y_i b) + 2\left(y_i^2 a^2 - 3y_i a + 3\right)\exp(y_i a)\end{array}\right]$$

$$a = \sqrt{3\mu_a\mu_s'}, \quad b = \sqrt{6\mu_s'^2 k_0^2 \alpha D_B T + 3\mu_a\mu_s'}, \quad x_i = \begin{cases} r_1^2 & i=1 \\ -2r_1 r_2 & i=2 \\ r_2^2 & i=3 \end{cases}, \text{ and}$$

$$y_i = \begin{cases} -2r_2 & i=1 \\ -(r_1+r_2) & i=2 \\ -2r_1 & i=3 \end{cases}.$$

$$K_s^2(\boldsymbol{r}) = \frac{2\beta C^2}{T}\left[A - \frac{1}{T}B\right] \tag{2.6}$$

This result enables leveraging high-density detection by camera sensors, as used by LSCI, with deep tissue probing by coherent point light sources, as used by DCT. Figure 2.1 illustrates the new need to account for variations in speckle contrast due to system parameters including SD separation and camera

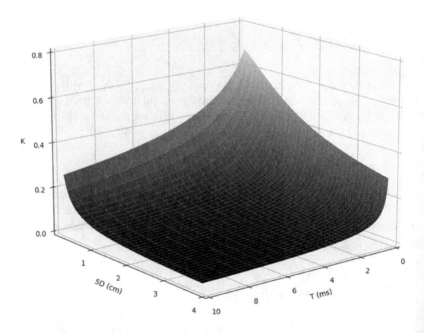

FIGURE 2.1   Diffuse speckle contrast theoretical variation with respect to SD separation and exposure time T system parameters over commonly used ranges.

exposure time. Each blood flow index, αD_B, is extracted through minimization of squared differences between the theoretical (K_{st}) and experimental (K_{se}) speckle contrasts by $\min\limits_{\alpha D_B}\left(K_{se}^2 - K_{st}^2\right)^2$ corresponding to a particular local region and SD separation. As shown throughout this chapter and the remainder of this book, there are a great many benefits to this design over using each technique individually.

INSTRUMENTATION

The scDCT hardware consists of light detection as modified from LSCI origins, light source generation as modified from DCT origins, and support apparatus. Figure 2.2 emphasizes the portability of scDCT and introduces the key parts. Note that specific hardware part numbers are representative with departures during applications mentioned explicitly. A highly sensitive electron-multiplying charge-coupled-device camera (EMCCD; Cascade 1K, Photometrics, AZ) or, most recently, a scientific complementary-metal-oxide-semiconductor camera (sCMOS; ORCA-Flash4.0, Hamamatsu) simultaneously detects photons across a dense pixel grid up to millions in number. An 830 or 785 nm long coherence length laser point source (CrystaLaser, NV) probes deeply into the tissue/sample. This combination eliminates the optical fiber bundles used by DCT while moving beyond superficial applications constraining LSCI.

Image reconstructions require boundary-measured BF information. Increasing the measurement number as well as spatial sampling distribution supports improving the ability to recover abnormalities in *a priori* unknown locations. Laser light is transferred via multimode optical fiber (FT200UMT, Thorlabs, NJ), through an achromatic lens for focusing (AC127-019-B, Thorlabs, NJ), to a galvo mirror (GVS002, Thorlabs, NJ) which in turn directs the laser beam to cover a desired ROI automatically and programmatically without physically contacting the tissue. A reflectance arrangement is used wherein the light source and detection reside at the same surface (i.e., as opposed to transmission). Polarizers in the source (LPNIRE050-B, Thorlabs, NJ) and detection (LPNIRE200-B, Thorlabs, NJ) beams are orthogonal to reduce reflected source light contributions from the sample surface. A zoom lens (Zoom 7000, Navitar, NY) focuses the re-emitted light into the camera detection sensor and completes full noncontact operation with flexibility toward different ROI's, subject scales, and overall translatability. A high-performance long-pass filter (84-762, EdmundOptics, NJ) eliminates ambient light and

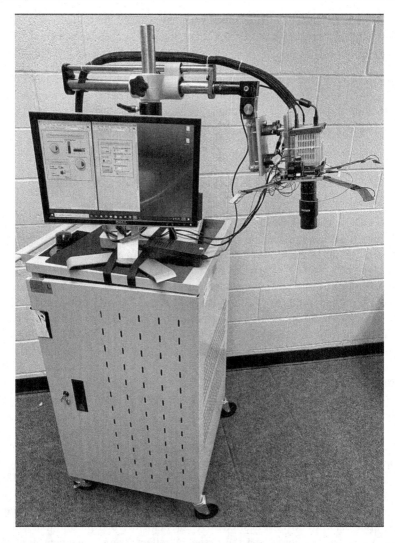

FIGURE 2.2   Base scDCT hardware system arranged on a mobile cart with articulating arm and including optional 3D photometric stereo technique (PST) apparatus.

other out-of-band light sources. An alternative arrangement combined a 785 nm long coherence length laser (CrystaLaser, NV) and long-pass filter (84-761, EdmundOptics, NJ).

An optional 3D photometric stereo technique (PST) can be enabled for fast and inexpensive recovery of the tissue surface contour [20]. Four 3D printed arms are installed at, and extend out from, the camera base. A neutral-white LED (SP-05-N4, Luxeon Star LEDs, AB, Canada) is attached to each arm for sequential illumination of the surface. The images from the different lighting vectors are then used to build up the geometry curvature.

BOUNDARY DATA COLLECTION

Any scDCT measurement procedure selects a tradeoff between SD spacing density and acquisition time. Upper limits on detector numbers are set by available pixels within a camera image, speckle size to pixel size ratio (Nyquist frequency), window sizes (local speckle pattern), and window averaging sizes (SNR). The hard upper limit on possible source positions is dictated by the galvo mirror scan step resolution. The soft upper limit is the associated added scan and exposure time involved with each new source position. For simplicity, it is common to set scans along a rectangular or circular grid such that we instead consider the implications of an additional row/column/radial line and their relative separations in accomplishing adequate ROI coverage. The temporal sampling resolution is then a consequence of acquiring the complete SD boundary data set based on SD density, camera frame rate (frames per second, FPS), and chosen exposure time.

Current iterations of scDCT have the camera integrate collected light intensity over a single exposure time typically ~5 ms. Sample flow dynamics cause a blurring in the otherwise static speckle pattern up to the exposure time when integration stops. Mismatching the exposure time with the time scale of flow dynamics can degrade the contrast sensitivity. See the literatures for further discussion [21].

With the collection sequence and setup appropriately enumerated, measurements are commenced by setting software parameters to manage the galvo mirror scan region and camera exposure time. The laser beam is oriented at the first scan position, takes an image, then continues through the remaining positions. This produces a data set for one representative time point. The process is repeated for averaging and/or the remainder of the study.

Once a complete set of images are stored (or on the fly), raw images undergo preprocessing and extraction of boundary flow data. Smear correction of raw images may be needed when using a frame transfer camera (i.e., using the Cascade 1K camera, but not the ORCA-Flash) and with sufficient light intensity levels during the exposure time used. Calculations using Equation 2.7

for the measured xth column and yth row, $i'(x,y)$, give the true pixel intensity $i(x,y)$ using frame transfer time t_{ft}, relative photoelectron generation efficiency η_{ft}, and number of pixel rows after frame readout n_p [22].

$$i(x,y) = i'(x,y) - \frac{\eta_{ft}t_{ft}}{n_pT} \sum_{n=1}^{y-1} i(x,n) \tag{2.7}$$

Other noise corrections to speckle contrast are that of dark and shot noise which are corrected through Equation 2.8 with light intensity I, dark current intensity I_D, dark corrected intensity $I_C = I - I_D$, and shot noise modeled by Poisson statistics $\sigma_s(I_C) = \sqrt{\mu(I_C)}$ [23,24].

$$K_s = \sqrt{\frac{\sigma^2(I) - \sigma^2(I_D) - \sigma_s^2(I_C)}{\mu^2(I_C)}} \tag{2.8}$$

A set of neighboring pixels constituting a local speckle pattern, and used in calculating an associated speckle contrast, is termed a window. The most common window size is a 7×7 pixel grid. A logical detector then refers to a set of neighboring windows that are averaged and correspond to a single SD separation. The most common logical detector size is a 3×3 window grid. Logical detectors are empirically selected to increase SNR, isolate target regions, and improve image reconstructions. Each calculated speckle contrast (logical detector) measurement could be used to track relative flow changes directly or, as is the current focus, be used for extracting a blood flow index by $\min_{\alpha D_B}\left(K_{se}^2 - K_{st}^2\right)^2$ for subsequent image reconstructions.

IMAGE RECONSTRUCTION

A flowchart of the imaging reconstruction process is given in Figure 2.3. The process is the same for scDCT as for its original beneficiary ncDCT, with two caveats. In scDCT, the boundary BF data is pulled from measured speckle contrasts associated with logical detectors. The SD separation for each logical detector is decided by the distance between the projected light spot on the tissue surface and the logical detector center. The handling of the reconstruction is equivalent to ncDCT thereafter and requires no modification. In other words, based on Figure 2.3, mesh generation, loading boundary data, and running the inverse model are as described in previous ncDCT publications. The intricacies involved are next summarized.

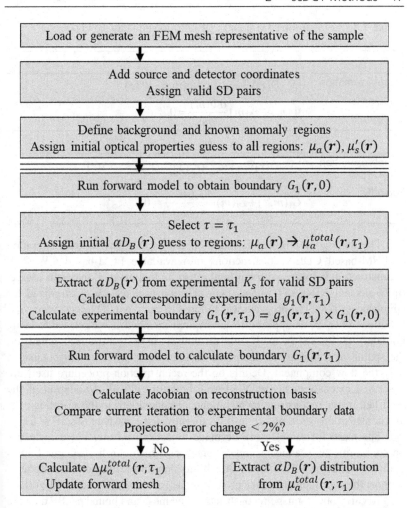

FIGURE 2.3 Flowchart depicting each step in the modified NIRFAST procedure including setting up the initial FEM mesh for the sample, incorporating measured boundary data, and the flow reconstruction procedure.

scDCT image reconstruction is a finite-element-based technique. The main theoretical underpinning is the continuous-wave (CW) photon diffusion equation (PDE) and CDE formal similarity. Their shared characteristics are evident by the view of Table 2.1 with continuous-wave source photon fluence rate $\Phi(\boldsymbol{r})$ (W/cm²), surface normal $\hat{n}$, and boundary position $\boldsymbol{m}$ [25–27].

TABLE 2.1   Comparison of PDE and CDE diffusion equations and boundary conditions

	PDE
P1 approximation	$\left[\nabla \cdot D(\boldsymbol{r})\nabla - v\mu_a(\boldsymbol{r})\right]\Phi(\boldsymbol{r}) = -vS(\boldsymbol{r})$
Robin BC	$\Phi(\boldsymbol{m}) + z_b(\boldsymbol{m})\dfrac{\partial\Phi(\boldsymbol{m})}{\partial\hat{n}} = 0$

	CDE
P1 approximation	$\left[\nabla \cdot D(\boldsymbol{r})\nabla - v\mu_a^{\text{total}}(\boldsymbol{r},\tau)\right]G_1(\boldsymbol{r},\tau) = -vS(\boldsymbol{r})$
Robin BC	$G_1(\boldsymbol{m},\tau) + z_b(\boldsymbol{m})\dfrac{\partial G_1(\boldsymbol{m},\tau)}{\partial\hat{n}} = 0$

The analogous problem formulation allows straightforward integration of FEM-based CDE reconstructions into available FEM-based CW PDE reconstruction software with minimal effort. In this manner, the open source, finite-element NIRFAST image reconstruction package was customized for BF purposes [28]. The program was tailored to carry out 3D tomographic distributions of μ_a^{total}, as opposed to its original intended μ_a target, while using boundary G_1 instead of Φ. If only differences in αD_B are unaccounted for, the spatial distribution and temporal fluctuations in μ_a^{total} are attributable to only the flow component. Details on the reconstruction procedure itself are described extensively in the NIRFAST literature [29,30]. Validation of the modified image reconstruction technique, considerations to the selection of τ, and contributions of μ_a and μ_s' to μ_a^{total} are provided elsewhere [28,31].

The mesh for samples can be made by several means. These include a photogrammetric scanner, simple geometries which can be generated automatically, and the PST method. Mesh creation and refinement may require support through supplementary programs such as ANSYS® and SolidWorks.

To carry out an image reconstruction, the mesh and boundary BF data are first fed into the modified NIRFAST program. Boundary BF data are used for calculating boundary G_1 data (see Figure 2.3). Other parameters must be supplied including SD separations (e.g., logical detector centers), measured or estimated optical property distributions for the tissue/phantom sample, correlation time, and source wavelength. Image reconstruction iterations run until < 2% projection error is found between the iteration-specific forward model solution and measured boundary G_1 data. Finally, the parameter of interest, αD_B, can be extracted from the definition of μ_a^{total}, providing a complete 3D BF distribution. This process is repeated for subsequent time points as necessary. rBF is found by normalizing to either an assumed healthy/baseline region αD_B (spatial) or to that identified by an important time point (temporal) or both.

REFERENCES

1. Fletcher, G.C., *Dynamic light scattering from collagen solutions. I. Translational diffusion coefficient and aggregation effects.* Biopolymers, 1976. **15**(11): p. 2201–2217.
2. Brown, W., *Dynamic Light Scattering: The Method and Some Applications.* 1993, New York: Clarendon.
3. Boas, D.A., L.E. Campbell, and A.G. Yodh, *Scattering and imaging with diffusing temporal field correlations.* Phys Rev Lett, 1995. **75**(9): p. 1855–1858.
4. Boas, D.A. and A.G. Yodh, *Spatially varying dynamical properties of turbid media probed with diffusing temporal light correlation.* J Opt Soc Am A-Opt Image Sci Vis, 1997. **14**(1): p. 192–215.
5. Schweiger, M., et al., *The finite element method for the propagation of light in scattering media: Boundary and source conditions.* Med Phys, 1995. **22**(11 Pt 1): p. 1779–1792.
6. Yu, G., et al., *Noninvasive monitoring of murine tumor blood flow during and after photodynamic therapy provides early assessment of therapeutic efficacy.* Clin Cancer Res, 2005. **11**(9): p. 3543–3552.
7. Durduran, T., *Non-invasive measurements of tissue hemodynamics with hybrid diffuse optical methods.* 2004, University of Pennsylvania, Dissertation.
8. Shang, Y., et al., *Diffuse optical monitoring of repeated cerebral ischemia in mice.* Opt Express, 2011. **19**(21): p. 20301–20315.
9. Buckley, E.M., et al., *Cerebral hemodynamics in preterm infants during positional intervention measured with diffuse correlation spectroscopy and transcranial Doppler ultrasound.* Opt Express, 2009. **17**(15): p. 12571–12581.
10. Roche-Labarbe, N., et al., *Noninvasive optical measures of CBV, StO(2), CBF index, and rCMRO(2) in human premature neonates' brains in the first six weeks of life.* Hum Brain Mapp, 2010. **31**(3): p. 341–352.
11. Kim, M.N., et al., *Noninvasive measurement of cerebral blood flow and blood oxygenation using near-infrared and diffuse correlation spectroscopies in critically brain-injured adults.* Neurocrit Care, 2010. **12**(2): p. 173–180.
12. Yu, G., et al., *Validation of diffuse correlation spectroscopy for muscle blood flow with concurrent arterial spin labeled perfusion MRI.* Opt Express, 2007. **15**(3): p. 1064–1075.
13. Zhou, C., et al., *Diffuse optical monitoring of hemodynamic changes in piglet brain with closed head injury.* J Biomed Opt, 2009. **14**(3): p. 034015.
14. Yu, G., *Diffuse Correlation Spectroscopy (DCS): A diagnostic tool for assessing tissue blood flow in vascular-related diseases and therapies.* Curr Med Imag Rev, 2012. **8**(3): p. 194–210.
15. Fercher, A.F. and J.D. Briers, *Flow visualization by means of single-exposure speckle photography.* Opt Commun, 1981. **37**(5): p. 326–330.
16. Bandyopadhyay, R., et al., *Speckle-visibility spectroscopy: A tool to study time-varying dynamics.* Rev Sci Instrum, 2005. **76**(9): p. 093110.
17. Boas, D.A. and A.K. Dunn, *Laser speckle contrast imaging in biomedical optics.* J Biomed Opt, 2010. **15**(1): p. 011109.

18. Yuan, S., et al., *Determination of optimal exposure time for imaging of blood flow changes with laser speckle contrast imaging*. Appl Opt, 2005. **44**(10): p. 1823–1830.

19. Huang, C., et al., *Noncontact 3-D speckle contrast diffuse correlation tomography of tissue blood flow distribution*. IEEE Trans Med Imaging, 2017. **36**(10): p. 2068–2076.

20. Mazdeyasna, S., et al., *Noncontact speckle contrast diffuse correlation tomography of blood flow distributions in tissues with arbitrary geometries*. J Biomed Opt, 2018. **23**(9): p. 1–9.

21. Liu, X., et al., *Simultaneous measurements of tissue blood flow and oxygenation using a wearable fiber-free optical sensor*. J Biomed Opt, 2021. **26**(1): p. 012705.

22. Ruyten, W., *Smear correction for frame transfer charge-coupled-device cameras*. Opt Lett, 1999. **24**(13): p. 878–880.

23. Huang, C., et al., *A wearable fiberless optical sensor for continuous monitoring of cerebral blood flow in mice*. IEEE J Sel Top Quantum Electron, 2019. **25**(1): p. 1–8.

24. Valdes, C.P., et al., *Speckle contrast optical spectroscopy, a non-invasive, diffuse optical method for measuring microvascular blood flow in tissue*. Biomed Opt Express, 2014. **5**(8): p. 2769–2784.

25. Durduran, T., et al., *Diffuse optics for tissue monitoring and tomography*. Rep Prog Phys, 2010. **73**(7): p. 076701.

26. Boas, D.A., *Diffuse photon probes of structural and dynamical properties of turbid media: Theory and biomedical applications*. 1996, University of Pennsylvania, Dissertation.

27. Zhou, C., *In-vivo optical imaging and spectroscopy of cerebral hemodynamics*. 2007, University of Pennsylvania, Dissertation.

28. Lin, Y., et al., *Three-dimensional flow contrast imaging of deep tissue using noncontact diffuse correlation tomography*. Appl Phys Lett, 2014. **104**(12): p. 121103.

29. Dehghani, H., et al., *Near infrared optical tomography using NIRFAST: Algorithm for numerical model and image reconstruction*. Commun Numer Methods Eng, 2008. **25**(6): p. 711–732.

30. Jermyn, M., et al., *Fast segmentation and high-quality three-dimensional volume mesh creation from medical images for diffuse optical tomography*. J Biomed Opt, 2013. **18**(8): p. 86007.

31. He, L., *Noncontact diffuse correlation tomography of breast tumor*, in *Biomedical Engineering*. 2015, University of Kentucky, Dissertation.

scDCT Validations and Applications

3

TESTING AND VALIDATING IN STANDARD PHANTOMS AND COMPUTER SIMULATIONS

Phantom tests and computer simulations are used to solidify expectations on performance regarding preliminary instrumental designs, base system modifications, and other unique circumstances surrounding specific applications. Both are beneficial due to their surrogate nature in creating an accurately representative physical (phantoms) or digital (computer simulations) target environment. The capability to adapt blood flow imaging (BFI) to applications often resides on initially accomplishing it with equivalent phantom or simulation testing. BFI as observed through these mechanisms can elucidate device characterization, proof-of-concept (PoC) designs, and optimize existing setups. Phantoms of known optical properties can also be used for calibration purposes and instrument adjustment. Two primary purposes emphasized in this section are system validation and SD range optimization. These are further described by the particular scDCT setup (e.g., PoC, full noncontact, iris) and the ROI for subjects of interest (e.g., mice, rats, piglets, humans). Herein, liquid phantoms were made from distilled water, India Ink (Black India, MA), and Intralipid (Fresenius Kabi, Sweden) and solid phantoms from carbon black, titanium dioxide, and silicone with procedures from the literature [1].

DOI: 10.1201/9781003246374-3

Computer simulations are in reference to image reconstructions generated using the modified NIRFAST package. Other supplemental software is also used such as for mesh refinement and extending measured sample boundary curvature to a solid model.

Proof-of-Concept Validation

Initial efforts to bring scDCT to fruition were based on a partial contact design in a reflectance setup [2]. This system is a simpler arrangement to the base version given in Figure 2.1. The 785 nm laser light was transferred by optical fiber into an optical switch (VX500, DiCon Fiberoptics, CA) instead of the galvo-scanning mirror. The four-channel, blocking switch sequentially conveyed light among four individual multimode fibers (FT200UMT, Thorlabs, NJ) arranged in a holder containing a central opening to the sample surface. The fibers were symmetrically located on the surface (ferrule tips) of a liquid phantom. This early system did not incorporate polarizers or a long-pass filter but gave valuable insights leading to the fully noncontact system. Importantly, it established a foundational ability to recover αD_B through speckle contrast K_s which could be directly incorporated into previously created 3D reconstruction algorithms for the noncontact DCT (ncDCT) device [3,4].

   Camera selection involves many factors and effects including costs, thermal and other noise levels, sensor size, frame rate, quantum efficiency, bit size (for analog to digital converter), and etaloning. When using the highly sensitive EMCCD camera, a key hindrance encountered in recovering quality raw images was the smearing effect introduced by the frame transfer process. Noticeable stripes occurred in bright areas along the direction of frame transfer over the set exposure time of 2 ms. This issue was resolved with a published smear correction algorithm [5]. After correction, measured speckle contrast correlated with theoretical predictions.

   To validate recovery of αD_B from K_s, simultaneous measurements of a homogeneous liquid phantom ($\mu_a^{785} = 0.05$ cm^{-1}, $\mu_s'^{785} = 7.0$ cm^{-1}, $\alpha D_B \cong 1 \times 10^{-8}$ cm^2/s) were compared between partial contact scDCT and standard DCS. For processing of partial contact scDCT images, 21 logical detectors were defined along a radial line across the center of the holder arrangement for each source. The SD range was from 1.0 to 2.0 cm with a 1.25 mm increment detector spacing. Noise-corrected speckle contrast calculations (Equation 2.8) were performed across 7×7 pixels, then averaged over 3×3 windows. This provided a single time point K_s corresponding to a single logical detector and was repeated for each remaining detector over 30 total frames. Corresponding flow indices were obtained by Equation 2.6. Good agreement (within 12%) was found between DCS and the partial contact scDCT.

An imaging test then utilized the new acquisition method for flow indices to achieve 3D image reconstruction. A cubical solid phantom with matching optical properties (μ_a, μ_s') to the liquid phantom but differing flow index (no flow) was placed in the center of the FOV, submerged 2 mm. The flow index was measured as 3 orders of magnitude less than the liquid phantom by DCS. The results from the modified NIRFAST software along with image display software ParaView (Kitware, NY) are shown in Figure 3.1. The normalized (to background) rBF is shown in Figure 3.1a and b. Identifying and characterizing the cubical heterogeneity as done by a half maximum contrast threshold is shown in Figure 3.1c and d along with simulated (ideal) results. Overall, the heterogeneity recovery was close to expectations with recovered center (0.0, −0.1, 25.0) compared to simulation recovered center (−0.1, 0.0, 24.0) and recovered rBF (0.38) to simulation recovered rBF (0.37). The side length was found to be 7.2 mm (actual: 7 mm).

Full Noncontact Validation and Optimal SD Separations for Larger ROI

To fully realize translatable potential while retaining reflectance-based subject interactions, a completely noncontact scDCT system was created. This is the representative base scDCT system shown in Figure 2.1, which includes the integrated PST as will be described shortly. The system attached the 830 nm laser and associated long-pass filter. The galvo-scanning mirror removes the static laser source placement, contact restriction, source fiber field-of-view (FOV) obstruction, and enables customizable automatic scanning patterns. It allows flexibility in source density and span for different ROI's and sampling depths. Cross polarizers reduce contributions of laser light reaching the detector due to reflections of the free space traveling, acutely incident beam from the sample surface. An added zoom lens provides manipulating the ROI in tandem with the scanned source patterns. A long-pass filter on the end of the zoom lens removes ambient light and is especially useful for clinical environments and shared areas.

In this study, a more elaborate scheme of tests provided extensive validation and insight on scDCT potential application performance. First, a container filled with homogeneous liquid phantom ($\mu_a^{785} = 0.05$ cm^{-1}, $\mu_s'^{785} = 8$ cm^{-1}, $\alpha D_B \cong 1 \times 10^{-8}$ cm^2/s) was measured by the scDCT placed directly overhead and with 5 ms exposure time. Equally spaced, within the ROI (80×80 mm^2), 9×9 source positions and 41×41 detector positions served as the total possible SD set. Averaging 4 frames per sample gave a sampling time of ~150 s. Data were processed as before with the same number of

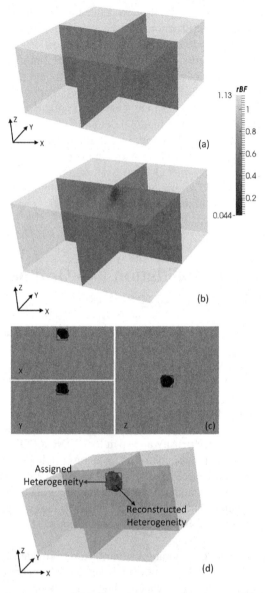

FIGURE 3.1 Reconstructed 3D flow contrasts with the partial contact scDCT. Cross-sectional views of the slab phantom are shown overlaid and against semi-transparent background (a) without heterogeneity and (b) with heterogeneity. (c) Orthogonal 2D cross sections through the center of the half maximum threshold extracted heterogeneity. Squares identify the bounds of the true heterogeneity positioning. (d) Half maximum extracted heterogeneity from computer simulation. (© 2015 Am. Assoc. Phys. Med from Ref. [2].)

pixels and windows grouped per detector. Selecting and analyzing four linear detector arrays it was found that those SD pairs in the range of 7–19 mm provided stable, precise flow data.

Using this optimal SD range for large ROI, a solid phantom test was undertaken similar in nature to that for the partial contact scDCT. In this case, two spherical solid (no flow) phantoms, optically matched to the background with ~7 mm diameters, were submerged with centers ~5 mm from the surface in separate areas. Results were comparable to the partial contact scDCT solid phantom test. A second phantom test involved a flow tube placed in either

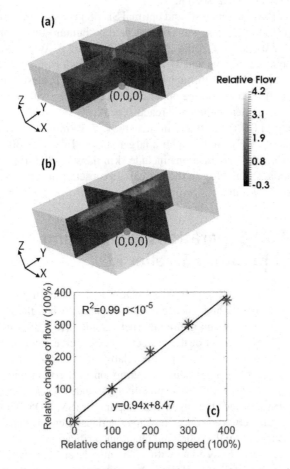

FIGURE 3.2 Reconstructed flow contrasts of the pump-connected tube at 20 ml/min pump speed. (a) and (b) give 2D cross-sectional results against semi-transparent background for the tube oriented along X and Y directions, respectively. (c) Relative pump flow as measured by scDCT accurately changing with relative pump speed increments. (© 2017 IEEE. Reprinted, with permission, from Ref. [6].)

of two orthogonal positions parallel to the liquid phantom surface and tube center at a depth of ~5 mm. The tube was transparent, cylindrical, connected to a peristaltic pump (HV-77201-60, Cole Parmer, IL), and filled with small solid phantom (as above) pieces to randomize particle motion of pumped liquid phantom. Figure 3.2 shows the reconstructed tube contrasts from both positions for a pump speed of 20 ml/min and agreement between rBF and the stepwise pump flow increase from 0 to 20 ml/min. Despite a complicated underlying relationship between the pumped flow (ml/min) and particle motion (cm²/s), it provided confidence in the sensitivity of fully noncontact scDCT to temporal changes in flow as well as spatial.

To verify the phantom tests and optimal SD range results were reasonable, a forearm cuff inflation test was done. A healthy human subject placed their forearm under the scDCT zoom lens. An ROI of 40 × 40 mm² was chosen with 5 × 5 source locations and 21 × 21 detector locations for ~30 s sampling time. Data were collected throughout a 4-min arterial occlusion induced by 230 mmHg cuff inflation on the upper arm. Results were within expectations and included an evident hyperemic reactive peak.

The optimal SD range found in this study (7–19 mm) is used by numerous subsequent scDCT studies with a larger subject ROI (e.g., 40 × 40 mm², 80 × 80 mm²). These applications include skin flaps [6–8], burn wounds [9], and human infant skull [10]. In the last case, a subset of the optimal SD range was selected (7–17 mm).

Optimal SD Separations When Using Iris Diaphragm for Smaller ROI

One of the primary selling points for scDCT is its touted translatability for subjects of varied sizes and surfaces. An issue arises when the ROI is compacted as is the case for small animals such as mice (10 × 10 mm²) and rats (20 × 20 mm²). Diffraction of the projected laser source can cause a disproportionately large intensity spot profile relative to the ROI size. As scDCT works with a point-like light source assumption rather than wide-field illumination such as with LSCI, this spreading precludes dense laser scanning. Adding a lever-actuated, half-open iris diaphragm (SM05D5, Thorlabs, NJ) in the source path removed source position overlaps to enable dense scanning options as evidenced by the following phantom verification.

The base scDCT system with 785 nm laser/filter with iris was characterized as follows [11]. A homogeneous liquid phantom ($\mu_a^{785} = 0.025 \, \text{cm}^{-1}$, $\mu_s'^{785} = 8 \, \text{cm}^{-1}$, $\alpha D_B \cong 1 \times 10^{-8} \, \text{cm}^2/\text{s}$) was measured using an ROI of 20 × 20 mm² to determine the SD operational range. A laser source at opposing corners (i.e., two sources) of the ROI was coupled with individual horizontal and vertical linear detector arrays (i.e., four total arrays). These arrays were each built from 24 logical detector positions at 0.5 mm steps to cover a 0.5–12 mm SD range from their respective source. Based on the stability of relative flow (normalized to their mean) and agreement between calculated and theoretical K_s^2 at each SD distance, it was determined that the optimal SD range for the system was between 2 and 8 mm.

This optimal SD range of the iris-based EMCCD scDCT facilitates measurements on smaller ROIs. A subset of this optimal SD range was subsequently chosen for mouse brain measurements. Specifically, with an ROI of 10×10 mm^2 the SD separations ranged from 2 to 6 mm [8].

Optimal SD Separations Using sCMOS Camera

To improve spatiotemporal resolution and remove the need for smear corrections, the EMCCD camera was replaced with the sCMOS camera (2048×2048 pixels, 30 Hz frame rate) for some scDCT studies. Explicit phantom testing was not performed, but for completeness it is mentioned here that the optimal SD range was based on excluding pairs with SNR level less than 3 or intensity saturation greater than 90% [12,13]. The application of this method includes 20×20 mm^2 and 40×40 mm^2 ROIs (with iris) on piglet brains (7–19 mm or 5–15 mm SD ranges) and 80×80 mm^2 on mastectomy skin flaps (4–19 mm SD range) [12,13].

Validation of PST for Arbitrary Surface Geometry Acquisition

Having an easy to use, built-in option for dealing with irregular geometries could positively impact many potential future applications of scDCT. A new PST method was proposed to overcome complexities with the original approach of using a photogrammetric scanner which required offline SD alignment and system co-registration [14]. With fast and inexpensive PST, surface geometries could be profited from through the existing cameras and minor scDCT system modifications.

The EMCCD scDCT system (without iris) with 830 nm laser and corresponding long-pass filter was upgraded with four 3D printed arms, each containing a neutral-white LED as shown in Figure 2.1. By sequencing the LEDs with a custom LabVIEW program (National Instruments, TX), a set of 2D images were obtained and represented the surface as shaded by the separate lighting vectors. From normal and FOV-based height maps the 3D sample boundary geometry was reconstructed and built into a full solid volume mesh [7]. The flow reconstruction procedure remained the same, except now the FEM representation was more realistic as to the true sample curvature. Locations of projected sources were determined by distinguishing areas of high localized intensity. These added features complete the base scDCT instrumentation setup from Figure 2.1.

Several tests sought to reconcile and repair irregular geometry influences on BFI by the scDCT system with PST. First, a patient having undergone mastectomy surgery was immediately imaged by PST to get a case study surface. The solid model was input into the flow-based FEM software. Based on future *in vivo* measurements, an ROI of 80 × 80 mm² was covered by a generated 9 × 9 source grid and 41 × 41 detector grid, of which SD in the range 7–19 mm were chosen, for computer simulation investigation. A spherical anomalous presence with a 5 mm radius and 2× flow contrast (with assumed background flow) was placed beneath the simulation sample surface. As expected, utilizing proper curvature resulted in much improved anomalous identification as compared with a typical flat surface model. With *a priori* geometry, the recovered sphere flow was 1.8× background (8.5% error) and ~5.7 mm radius (~14% error). With assumed flat surface, recovered sphere flow was 1.27× and the shape was too severely distorted to apply an interpretation of radius. A forearm arterial occlusion by cuff inflation test was done, repeating the protocol from the scDCT forearm study described earlier, with 220 mmHg for ~160 s and sampling time of ~40 s. Physiological changes were as expected with a global BF decrease on cuff inflation and hyperemic response and return to normal after cuff release. This response with PST-measured forearm curvature is expected to better represent the true response as compared to the previous case with a flat model.

Validation on Human Infant Skulls

A preliminary study in preterm infants employed the base EMCCD scDCT system (without iris) with the 785 nm/filter combination. Prior to *in vivo* analysis of neonatal brain, an infant-head-simulating phantom was created for optimizing scDCT flow recovery in a mock setup. An infant skull (31 weeks after fertilization; Evolution Store) was held in an inverted orientation and filled with liquid phantom. The liquid phantom optical properties were matched to reported values for the infant brain at 788 nm. To evaluate scDCT ability to penetrate the skull (~2 mm thick at ROI) and accurately detect phantom CBF, the outcome was validated against DCS. The scDCT ROI was directed upon the foramen magnum and covered an area of 30 × 30 mm². The SD range

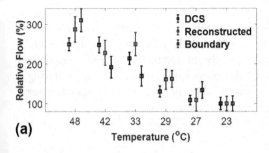 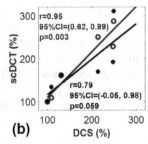

FIGURE 3.3 Flow recovery through infant skull with scDCT and DCS. (a) Relative flow changes (mean ± standard deviation) for 5-min stepwise decreases in liquid phantom temperature. (b) scDCT relative flow changes from the boundary (solid dots) or as reconstructed (circles) compared with DCS. Fitting lines are by linear regression. Pearson's correlation coefficient is denoted by *r*. (© Institute of Physics and Engineering in Medicine. Reproduced by permission of IOP Publishing Ltd from Ref. [10] (DOI: 10.1088/1361-6560/abc5a7). All rights reserved.)

subset (of optimal) was from 7 to 17 mm for scDCT and a single DCS SD of 15 mm. Figure 3.3a graphs the relative flow (to room temperature) sensitivity of scDCT to changes in Intralipid particle motion associated with a stepwise temperature drop to room temperature. With the SD separations used, approximately half of such distances expected to highlight common penetration depth, and the skull thickness, these findings show scDCT can detect flow through infant skull. Figure 3.3b (circles) supports that these quantifications represent actual flow with significance between scDCT at 5 mm depth and DCS.

APPLICATION 1: IMAGING OF BRAINS

The brain functions in a highly regulated environment where disturbances can lead to or indicate disease, damage, or other adverse conditions. CBF is a crucial biomarker commonly associated with alterations in the brain micro-environment and hemodynamic revisions. The longitudinal and continuous monitoring of CBF can open pathways to understanding developments such as acute or chronic hemorrhaging, ischemia, and injury. Measuring CBF has unique challenges, notably the existence of, and need to pass through, the skull layer. The noninvasive NIR optical nature of scDCT is well-equipped for this capacity and established diffuse NIR methods have already had success. In this section, we explore four applications which customized and tested scDCT for BFI of deep brain CBF. The first and second studies updated scDCT for imaging CBF through the small available surface region of mouse and rat heads, greatly benefiting small animal studies often incorporated into early research plans [11]. The third study showed the updated scDCT device's capability for assessing BFI of CBF in neonatal piglets, known for having a high resemblance to human neonates [12]. The fourth study probed preterm infant CBF with base scDCT instrumentation through a transparent incubator wall without causing discomfort to this highly vulnerable and fragile population [10].

With rodents continuing to comprise 95% of the animals used in contemporary biomedical research it was imperative to venture into this territory with the scDCT system. As mice are of generally higher availability than most species, they also tend to be the easiest to quickly test hypotheses in large groups and sit on substantial research findings for comparisons. Unfortunately, they also tend to be unreliable in rewarding insights intuitively translatable to humans. Neonate rats, on the other hand, closely resemble human neonates. Mice and rats were accordingly the first animal models to be tested in detail with scDCT. The system applied was the base scDCT with iris.

Assessing CBF in a Mouse

One mouse was placed in a stereotaxic frame and anesthetized prior to head hair removal. The scalp and skull were otherwise undisturbed. The iris diaphragm was adjusted to compensate for the small 10×10 mm^2 ROI of the mouse head. Common carotid artery (CCA) ligations were then performed in the following protocol: (1) baseline, (2) ipsilateral (left) ligation, (3) bilateral ligation, (4) right ligation release. Reconstructions of BFI used 5×5 source (0.025 Hz) and 21×21 detector grids with SD separations between 2 and 6 mm (subset of optimal SD for small ROI). Results are shown in Figure 3.4. Trends in rCBF followed expectations, decreasing in trend with corresponding ligation applications, increasing with ligation release, and doing so according to depth (skull or deep brain) and region (left or right). Left CCA occlusion alone induced a 40% rCBF drop while bilateral arterial occlusions induced a 75% rCBF drop as measured by scDCT. This success on mice affirmed application of the iris-based scDCT to an animal at the low end of subject size scaling.

Assessing CBF in Rats

Ten rats (adult male Sprague-Dawley of 2–3 months age) were also imaged by scDCT in a separate study. After shaving and cleaning their head, nine

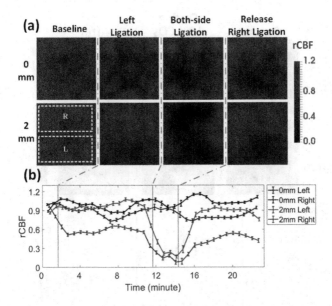

FIGURE 3.4 3D rCBF distributions in a mouse with CCA ligations. (a) Cross-sectional views at 0 mm (scalp) and 2 mm (cortex) depths of rCBF as reconstructed by scDCT. R and L boxes denote right and left hemispheres, respectively. (b) Time-course changes in rCBF at the two depths and hemispheres throughout ligation phases. Error bars represent standard (spatial) deviations. (© 2018 IEEE. Reprinted, with permission, from Ref. [8].)

rats went through two protocols: a 10% CO_2/90% O_2 inhalation procedure; transient unilateral and bilateral ligations of CCA. CO_2 is an often-used vascular dilator for inducing CBF increase. CCA ligation decreases CBF in the main consumer (left/right) hemisphere. The tenth rat was imaged intermittently over 14 days before, during, and after stroke by transient middle cerebral artery occlusion (MCAO). Images for 3D reconstruction were processed with 5 × 5 sources and 21 × 21 detectors.

The results from a ~10 min application of the CO_2/O_2 mixture can be seen in Figure 3.5. Baseline and recovery periods are clearly distinguishable from the induced CBF increase. The average CBF response from the healthy rat group was similar to that found in literatures. From this, it was verified that the modified scDCT for rat brain detected CBF changes due to the normal CO_2 inhalation technique. Figure 3.6 depicts 3D reconstructed rCBF throughout CCA ligation with an emphasis on the separable superficial scalp tissue and deep cortex tissue response. Notably, significant differences calculated by paired t-tests were found between the two hemispheres during unilateral

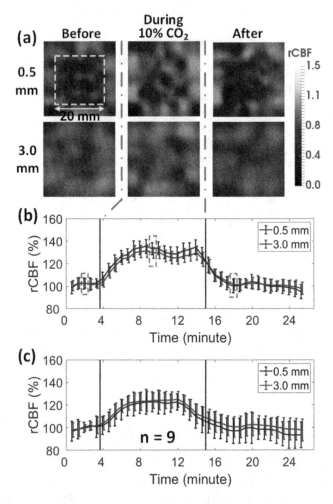

FIGURE 3.5 Responses in rat rCBF by 3D scDCT reconstructions to 10% CO_2/90% O_2 inhalation. (a) Representative rat response (rat #5) throughout each phase of the inhalation procedure. Cross-sectional views are shown for 0.5 mm (scalp/skull) and 3.0 mm depths (cerebral cortex). (b) Corresponding inhalation phase responses over time for each depth in (a) where error bars are standard deviations. (c) The response of all rats over time throughout inhalation where error bars are standard error over the 9 rats. (Reprinted from Ref. [11], © 2019, with permission from Elsevier.)

ligation and recovery, between the two layers, and in both layers of the ligated hemisphere(s) for all ligation stages. These results support continuous scDCT imaging to detect regional cerebral blood flow variations in larger rodent species.

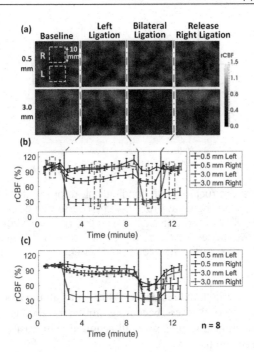

FIGURE 3.6 Responses in rat rCBF by 3D scDCT reconstructions to unilateral and bilateral CCA ligations. (a) Representative rat response (rat #3) throughout each ligation phase. Cross-sectional views are shown for 0.5 mm (scalp/skull) and 3.0 mm depths (cerebral cortex). For subsequent graphs, averaged rCBF for each hemisphere is within the area (10×10 mm^2) bounded by dashed square. (b) Responses over time for each depth in (a) where error bars are standard deviations. The dashed boxes in this case marks where the images were pulled from for (a). (c) The response of all rats over time where error bars are standard error over the 8 rats (one excluded due to surgical complication). (Reprinted from Ref. [11], © 2019, with permission from Elsevier.)

For the two protocols just mentioned, this study also reported on a 2D CBF mapping technique. The advantage of such a technique is the rapid feedback of flow alterations (seconds versus minutes). The conclusions generally agreed with the full 3D reconstructions with some expected diminishing in quality and accuracy. This reduction can be attributed to partial volume effects that over-/under-estimate the deep tissue flow. Compared to superficial wide-field illumination techniques, the 2D scDCT mapping offers a quick look into deep tissue flow activity.

A longitudinal test fills out this study by monitoring rCBF changes before, during, and for 14 days after an MCAO-induced stroke of the left hemisphere in the final rat (#10). From Figure 3.7, the ipsilateral side was affected most

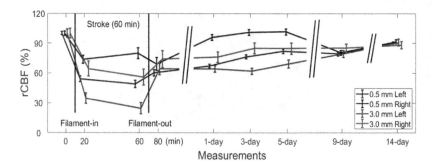

FIGURE 3.7 Reconstructed rCBF longitudinal response to MCAO-induced stroke of the left hemisphere (rat #10). The time-course changes are conveyed for 0.5 mm (scalp/skull) and 3.0 mm depths (cerebral cortex) of both hemispheres until 14 days post stroke. Error bars represent standard deviations. (Reprinted from Ref. [11], © 2019, with permission from Elsevier.)

by the filament insertion and residence. After removal, the blood flow tended toward baseline with all layers becoming similar by the end. These, along with the aforementioned results, were all acquired noninvasively and without invasive scalp retraction.

Neonatal Piglet CBF Recovery

CBF in newborn infants is affected by perinatal diseases such as intraventricular hemorrhage (IVH) and ischemia/asphyxia. Studies with PET, CT, and MRI have identified associated regional CBF decreases with these conditions and a relationship between positive treatment responses and CBF [15–20]. However, their bulky equipment, cost, and/or radioactive tracers inhibit these modalities in achieving clinical bedside usage. The next study recognized that scDCT could offer an alternative and safe bedside technology for continuous infant brain imaging in neonatal intensive care units (NICU). It thus proceeded to test recreating and detecting ischemia/asphyxia and IVH effects on two neonatal piglets (analogous to human neonates [21–28]) by scDCT. The piglet brain also represents a mid-sized subject application of scDCT.

The sCMOS scDCT system with a lever-actuated iris was used for this study. Although the ROI was not as small as in the rat head, the iris here instead confined the laser spot size to allow intensity adjustment for the newly installed camera. The scDCT was oriented over the intact scalp (piglet #1) or exposed skull (left skull of piglet #2). Four frames were averaged by scDCT per 5 × 5 source grid. A 21 × 21 detector grid was used for both piglets, but an individual ROI of 40 × 40 mm² for piglet #1 and of 20 × 20 mm² for piglet #2. A contact DCS probe at a non-interfering distance from scDCT, for comparisons, and an intracranial pressure (ICP) measurement device (ICP Express Monitor, 82-6634H8, Codman) were placed at/into the right lateral ventricle. Heart rate (HR), respiration rate (RR), and arterial blood oxygen saturation (SaO₂) were also monitored by multi-channel Respirator-Oximeter (8400, Smiths Medical) secured to the tongue or ear in a 30 min rotation.

Each piglet of two female neonatal Yorkshire piglets (postnatal day 9) was exposed to asphyxia and either transient global ischemia (piglet #1) or IVH (piglet #2). The ischemia protocol consisted of baseline, right CCA

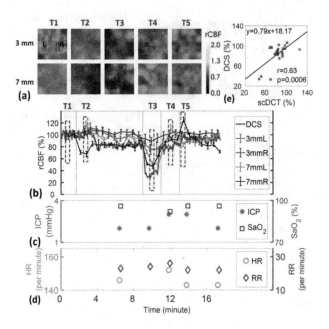

FIGURE 3.8 Piglet #1 transient unilateral and global ischemia results. (a) 3D reconstructed rCBF by scDCT at cross-sectional depths of 3 mm (scalp/skull) and 7 mm (cortex) for each phase. T1: baseline; T2: right CCA ligation; T3: bilateral ligation; T4: left ligation release; T5: bilateral ligation release and recovery. rCBF separately averaged for left and right hemispheres were over their respective areas bounded by 20 × 10 mm² dashed squares. (b) Concurrent DCS and scDCT time-course changes in rCBF with dashed boxes representing scDCT images pulled for (a). Error bars are standard deviations within each region. (c) and (d) are time-course changes in the measurements from ICP and HR, RR, and SaO₂. (e) Linear regression between cortex rCBF as obtained by scDCT and DCS. (© 2020 Wiley-VCH GmbH from Ref. [12].)

occlusion, bilateral CCA occlusion (global ischemia), left occlusion release, and bilateral occlusion release. The asphyxia protocol graduated an O_2/CO_2 mixture from 100% O_2/0% CO_2 to 5% O_2/95% CO_2 over an 8 min duration. The IVH protocol consisted of baseline, injection of 1 ml heparinized autologous blood over 5 min (IVH), saline injections to maintain elevated ICP for ~60 min, and 30 min recovery. The asphyxia protocol was performed after recovery from the previous protocol (ischemia or IVH) and immediately prior to euthanasia for both piglets. Results for piglet #1 ischemia (Figure 3.8), piglet #2 IVH (Figure 3.9), and piglet #2 asphyxia (Figure 3.10) are provided.

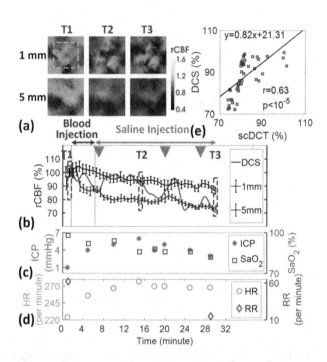

FIGURE 3.9   Piglet #2 IVH results. (a) 3D reconstructed rCBF by scDCT at cross-sectional depths of 1 mm (skull) and 5 mm (cortex) for each phase. T1: baseline; T2: IVH; T3: end of measurement. rCBF was averaged over the area bounded by the 15 × 15 mm² dashed square. (b) Concurrent DCS and scDCT time-course changes in rCBF with dashed boxes representing scDCT images pulled for (a). Error bars are standard deviations within each region. Triangles mark saline injections. (c) and (d) are time-course changes in the measurements from ICP and HR, RR, and SaO₂. (e) Linear regression between cortex rCBF as obtained by scDCT and DCS. (© 2020 Wiley-VCH GmbH from Ref. [12].)

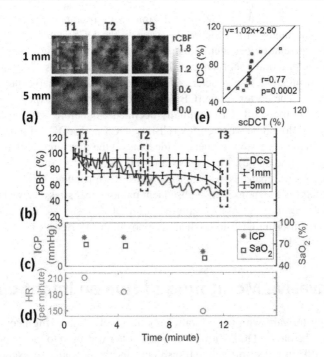

FIGURE 3.10 Piglet #2 asphyxia results. (a) 3D reconstructed rCBF by scDCT at cross-sectional depths of 1 mm (skull) and 5 mm (cortex) for each phase. T1: baseline; T2: asphyxia; T3: end of measurement. rCBF was averaged over the area bounded by the 15 × 15 mm² dashed square. (b) Concurrent DCS and scDCT time-course changes in rCBF with dashed boxes representing scDCT images pulled for (a). Error bars are standard deviations within each region. (c) and (d) are time-course changes in the measurements from ICP and HR, RR, and SaO₂. (e) Linear regression between cortex rCBF as obtained by scDCT and DCS. (© 2020 Wiley-VCH GmbH from Ref. [12].)

The rCBF response by scDCT from piglet #1 (Figure 3.8b) to induced ischemic matched expectations with respect to the different layers and hemispheres. For example, only the right cortex rCBF noticeably decreased with right CCA ligation, the left cortex rCBF decreased in coordination with the added left CCA ligation, and some hyperemic responses occurred after ligation release. Significant correlations were observed between the scDCT and DCS measurements (Figure 3.8e). Brief ICP elevations coincided with rCBF hyperemic response and increased blood volume (Figure 3.8c). The rCBF response by scDCT from piglet #2 (Figure 3.9b) to IVH also met expectations.

Both layers decreased gradually over time with prolonged high-pressure exposure. Although some instability existed in the DCS signal, there was a significant correlation between it and scDCT cortex rCBF (Figure 3.9e). ICP increased in temporal congruence with IVH/saline procedures. The asphyxia response for piglet #2 as characterized by scDCT rCBF matched expectations (Figure 3.8b). The cortex tissue rCBF declined with hypoxic exposure and to a lesser extent the skull rCBF mutually declined. As before, significant correlations were found between cortex rCBF by DCS and scDCT (Figure 3.8e). Hypoxic stress also visibly influenced (decreased) the measured ICP, HR, and SaO_2 (Figure 3.8c and d). The asphyxia results for piglet #1 (not shown) presented some discrepancies with that of piglet #2, attributed to intact scalp, individual animal response, and subsequent pathological stress. Nevertheless, these overall findings gave final approval of scDCT in successfully and continuously monitoring CBF for ischemia/hypoxia and IVH with high bedside NICU potential.

Noninvasive Monitoring of Human Infant CBF

Continuing forward with the notion of noninvasively monitoring neonates in the NICU, the base scDCT with 785 nm laser/filter was used to image preterm infants' CBF in-place through a transparent incubator wall. To exclude the influence of the incubator, an adult human forearm was placed inside and measured using a 40 × 40 mm^2 ROI, 5 × 5 source grid, and 21 × 21 detector grid. A cuff inflation protocol, in line with previous usages, returned rBF corresponding to a baseline, arterial occlusion, and recovery of 4 min each. This was done on the forearm through free space and through the incubator wall with good agreement found between the two (i.e., negligible wall influence).

With the incubator wall contributions determined negligible, two extremely preterm infants (#1: male, born at 25 1/7 weeks gestation, 815 g birth weight; #2: male, born at 26 5/7 weeks gestation, 995 g birth weight) were imaged (CBF) through it. CBF BFI was taken on infant #1 at 1, 2, and 3 weeks of life for longitudinal monitoring. Infant #2 response to patent ductus arteriosus (PDA) pharmacotherapy was based on imaging before and after 2 h indomethacin treatment. Each measurement collected data for ~5 min duration. Reconstructed flow distributions for both infants are provided in Figure 3.11. As expected, one-way ANOVA indicated no significant differences between rCBF across the 3 weeks for infant #1. For infant #2, a significant increase, as calculated by Student's t-test, in rCBF occurred after PDA pharmacotherapy. Blood pressure was found to increase as well. Importantly, peripheral oxygen saturation measured by standard finger pulse oximetry did not significantly change.

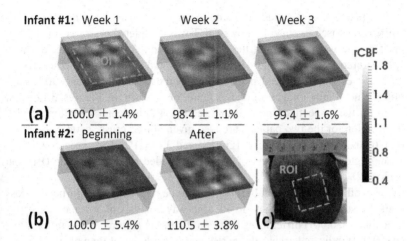

FIGURE 3.11 Measures of rCBF by scDCT in two extremely preterm infants. (a) Resulting rCBF at 6 mm below the scalp for infant #1, exhibiting only slight variations over the span of 3 weeks. (b) Pharmacotherapeutic effect on rCBF (6 mm below scalp) observed in infant #2. The rCBF elevation was significant and correlated with blood pressure increase. (c) The ROI as visible in a plain photograph which is the same as indicated in (a). (© Institute of Physics and Engineering in Medicine. Reproduced by permission of IOP Publishing Ltd from Ref. [10] (DOI: 10.1088/1361-6560/abc5a7). All rights reserved.)

APPLICATION 2: IMAGING OF BURN/WOUND TISSUES

A condition where compromised tissue is not fit for direct contact is with burn wounds. Noncontact BFI options can overcome concerns related to contaminating and infecting the wound area and otherwise introducing adverse negative effects. This asset is more pronounced when assessing injuries on the battlefield and at combat surgical hospitals with constrained resources and specialized objectives. Deficiencies in contemporary BFI modalities such as Doppler ultrasound, LSCI, PET, SPECT, XeCT, and PCT also exist within the context of the proposed problem. Size, expense, penetration depth, and ionizing radiation are either not conducive to portability and ubiquity or introduce unsafe side effects. A preliminary study into scDCT performance as a possible resolution is now examined [9].

The 830 nm laser/long-pass filter scDCT (Figure 2.1) measured rBF distributions in full-thickness burn wounds for two patients under anesthesia.

A third patient was excluded due to motion artifacts arising from measurements while in a conscious state. BFI by scDCT was completed prior to excising and grafting the wound while in the operating room. An ROI of 80 × 80 mm^2 was covered by a 9 × 9 and 41 × 41 sources and detector grid, respectively. Two frames were averaged per sample (source position) giving ~90 s sampling time. Plain images of the wounds are shown in Figures 3.12a and 3.13a for the respective first (P1) and second (P2) patients. Variations in rBF distribution for the two patients are also given at four cross sections (depth of 0, 3, 6, and 9 mm) in Figures 3.12b (P1) and 3.13b (P2). Categorization of the tissue regions and their level of rBF are provided in Figures 3.12c (P1) and 3.13c (P2).

In both patients, rBF in burn/wound tissue regions were significantly less than the surrounding tissues. In P1, one-way analysis of variance gave significant ($p < 0.001$) differences between the tissue regions, while in P2, t-test ($p < 0.001$) indicated eschar tissue rBF significantly lower than nearby tissues. The greater deviations in P2 damaged tissue rBF from normal also supplied a better match with the plain image. These results hold promise for scDCT in safely retrieving tissue burn and wound status and thereby improving site assessment without interfering with the damaged site.

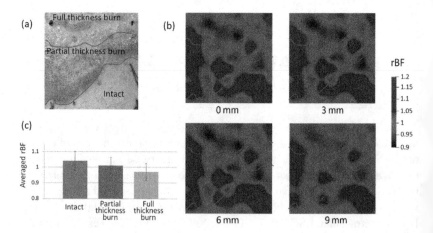

FIGURE 3.12 Reconstructed blood flow distributions in burn wound patient #1. (a) Plain photograph of the burn site. (b) Cross-sectional views of reconstructed rBF at depths of 0, 3, 6, and 9 mm. (c) Average rBF across different tissue volumes with standard deviation error bars. (Reprinted from Ref. [9] by permission of Oxford University Press.)

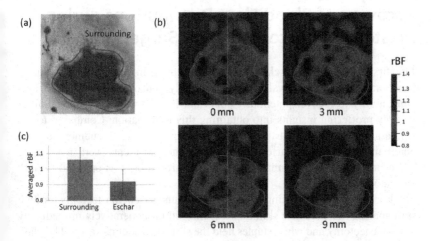

FIGURE 3.13 Reconstructed blood flow distributions in burn wound patient #2. (a) Plain photograph of the burn site. (b) Cross-sectional views of reconstructed rBF at depths of 0, 3, 6, and 9 mm. (c) Average rBF across different tissue volumes with standard deviation error bars. (Reprinted from Ref. [9] by permission of Oxford University Press.)

APPLICATION 3: IMAGING OF MASTECTOMY SKIN FLAPS

Mastectomy is conducted in many women with symptomatic breast cancer. Detection of postoperative complications including skin flap ischemia and necrosis is vital to achieving maximum restoration success. The delicate nature of the site and technological limitations have limited the role of imaging instrumentation in assisting the medical professional. The primary technology in this space, fluorescence angiography (FA), can be invasive and costly. Clinical judgment by FA (e.g., indocyanine green (ICG) angiography) is the standard for assessing skin flap viability [29]. A factor limiting any imaging instrument in this realm is insufficient penetration into the full mastectomy skin flap thickness. This imposes difficulties with generally superficially probing technologies such as standard LSCI [30]. These circumstances suggest a deficiency that the scDCT instrumentation can handle by noninvasively characterizing revascularization through BFI and providing a spatially descriptive picture of necrotic or otherwise failing areas.

Recovery of Blood Flow Distributions during Mastectomy Reconstruction Surgery

The first application of scDCT to mastectomy skin flaps was a mutually beneficial study. That is, validation of the fully noncontact base scDCT system's ability to recover potentially abnormal BF distributions while simultaneously proffering insights into obtaining this information. Compared to the validation tests for this particular system setup (Figure 3.2, submerged solid phantoms, and forearm arterial occlusion), measurements were now being taken in a clinical environment with ambient room light as opposed to dark laboratory conditions [6].

For this case study, a single female subject underwent mastectomy through skin-sparing incision. The skin flap was imaged intraoperatively immediately after mastectomy and while staples held the skin closed temporarily. The plain image of the incision site to be measured can be seen in Figure 3.14a. The selection of ROI, source locations, detector locations, and flow processing matched the burn wound study. Two frames were averaged per source with ~80 s sampling time. Inspection of the reconstructed 3D BF distributions in Figure 3.14b and c shows a noticeable BF drop around the edge of the mastectomy skin flap. Comparing this functional information to the plain image (Figure 3.14a), the edges generally coincide and reveal a similar shape and orientation. This result promoted confidence in scDCT for supplying BF distributions which may be evaluated to, for example, guide intraoperative procedures in removing compromised tissues and reducing the incidence of flap necrosis.

Incorporating Measured Surface Curvatures by PST

Despite the previous success, the surface geometry of mastectomy skin flap areas is not expected to match ideal or phantom surface curvatures. With the PST-capable scDCT instrument, measuring BF distributions of mastectomy skin flaps was revisited to see if the results could be improved [7]. A female patient was measured for irregular tissue boundary by PST and BFI by scDCT. In the patient, data was collected after nipple-sparing mastectomy and staples had temporarily closed the incision site (Figure 3.15). An 80×80 mm^2 ROI, 9×9 source grid, and 41×41 detector grid followed suit to previous studies with a subject-specific mesh based on information from PST. scDCT averaged 2 frames per sample with a sampling time of ~2 min. Total measurement time (PST + scDCT) was thus kept under 2 min with minimal clinical impact.

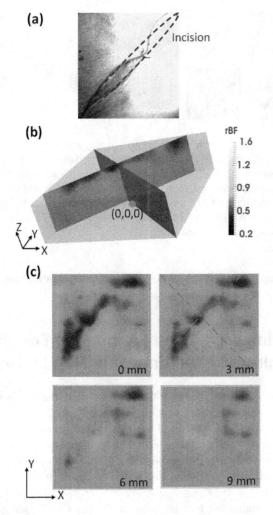

FIGURE 3.14 Case study results with scDCT for mastectomy skin flap intraoperative imaging. (a) Plain photograph of mastectomy skin flap site. (b) Reconstructed flow distributions along the skin flap site as cross sections against semi-transparent background. (c) Cross sections through the X-Y plane at depths of 0, 3, 6, and 9 mm. Dashed lines coincide with cross sections shown in (b). (© 2017 IEEE. Reprinted, with permission, from Ref. [6].)

High and low spatial variations in rBF are visible at several cross-sectional depths (0, 3, 6, and 9 mm) from Figure 3.15g. In the results, the rBF is relative to the mean BF within the volume. Importantly, scDCT noninvasively supplied

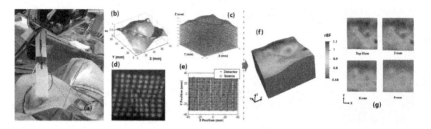

FIGURE 3.15 Intraoperative mastectomy skin flap imaging revisited with scDCT, now including PST capability on human patient. (a) Setup of scDCT with PST on the imaging site. (b) Curvature of the tissue surface obtained by PST and subsequent (c) mesh. (d) Projected sources on the tissue surface and (e) the complete source and detector grid. (f) and (g) show the 3D flow distribution and cross-sectional views at 0, 3, 6, and 9 mm depths, respectively. (Reproduced from Ref. [7] under Creative Commons BY 4.0.)

information up to 9 mm depth, close to average postoperative mastectomy skin flap thicknesses of ~10 mm [31].

Comparing scDCT and FA Imaging Equivalence for Judging Mastectomy Skin Flap Status

From the previous study, it was anticipated that with the combined PST and scDCT instrumentation, partial or full-thickness mastectomy skin flap viability may be assessed rapidly and harmlessly for optimal breast reconstruction outcomes. To better scrutinize this early optimistic conclusion, a more comprehensive study was done comparing scDCT (with PST) to the imaging standard for viability judgment, FA, in human mastectomy skin flap reconstructions [13].

Eleven human adult females aged 31–66 years and undergoing nipple-sparing or skin-sparing mastectomy followed by single-stage reconstruction (direct to implant reconstruction) or two-stage reconstruction (tissue expander followed by implant reconstruction) were recruited for this expanded study. The EMCCD camera component of scDCT was replaced with the sCMOS camera. For PST, NIR illumination LEDs (850 nm, Luxeon Star LEDs, AB, Canada) capable of passing through the long-pass filter were also substituted in. BFI was measured for each patient by scDCT with an 80×80 mm^2 ROI, 9×9 source grid, 41×41 detector grid, and 2 ms exposure time. scDCT rCBF images were co-registered with FA images (SPY-PHI, Novadaq/Stryker) taken from the same region and temporal proximity.

Representative results from both imaging techniques with individually segmented contours regarding a single patient (#11) are shown in Figure 3.16. The scDCT 3D rBF distributions and segmented regions convey more complete variations in tissue BF status by resolving with depth in comparison to the 2D mapping of FA. Full analysis of scDCT and FA equivalence in BF assessment was quantified by calculating Pearson's correlations for 8 contours in 4 different areas (10×10 mm^2 to 40×40 mm^2) over 11 patients. Significance, considered as $p < 0.05$, was found for all contours in the 10×10 mm^2 area ($r \geq 0.78$, $p < 0.004$) and 5 of the 8 contours in the 20×20 mm^2 area ($r \geq 0.73$, $p < 0.02$).

An argument can be made that scDCT has a vital role in future non-invasive imaging for mastectomy skin flap recovery judgment. Although more patients need to be recruited for greater statistical power, there was an

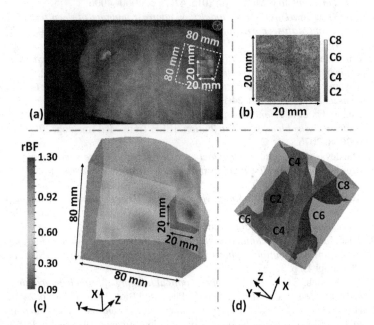

FIGURE 3.16 Comparison of scDCT and SPY-PHI imaging results for patient #11. (a) The SPY-PHI image with the following overlays: scDCT 80×80 mm^2 ROI; 20×20 mm^2 area situated on lowest BF as detected by scDCT; ellipsis identifying high-intensity perfusion artifacts. (b) Segmentation of the 20×20 mm^2 square into 8 regions/contours based on perfusion levels. (c) The BF distribution from scDCT presented as a 3D view with selected $20 \times 20 \times 20$ mm^3 ischemic area based on the lowest BF. (d) Segmented volumes/contours based on BF levels within the cube of (c) where only 4 of 8 are shown to promote distinguishing visually. (Reproduced with permission from Ref. [13] (DOI: 10.1097/PRS.0000000000009333).)

agreement between scDCT and FA in the most localized low BF areas. These areas are often the most important during reconstruction surgery. This was accomplished without dye injections and with relatively inexpensive, noncontact instrumentation.

REFERENCES

1. Choe, R., *Diffuse optical tomography and spectroscopy of breast cancer and fetal brain.* 2005, University of Pennsylvania, Dissertation.
2. Huang, C., et al., *Speckle contrast diffuse correlation tomography of complex turbid medium flow.* Med Phys, 2015. **42**(7): p. 4000–4006.
3. Lin, Y., et al., *Three-dimensional flow contrast imaging of deep tissue using noncontact diffuse correlation tomography.* Appl Phys Lett, 2014. **104**(12): p. 121103.
4. Lin, Y., et al., *Noncontact diffuse correlation spectroscopy for noninvasive deep tissue blood flow measurement.* J Biomed Opt, 2012. **17**(1): p. 010502.
5. Ruyten, W., *Smear correction for frame transfer charge-coupled-device cameras.* Opt Lett, 1999. **24**(13): p. 878–880.
6. Huang, C., et al., *Noncontact 3-D speckle contrast diffuse correlation tomography of tissue blood flow distribution.* IEEE Trans Med Imaging, 2017. **36**(10): p. 2068–2076.
7. Mazdeyasna, S., et al., *Noncontact speckle contrast diffuse correlation tomography of blood flow distributions in tissues with arbitrary geometries.* J Biomed Opt, 2018. **23**(9): p. 1–9.
8. Mazdeyasna, S., et al., *Noninvasive noncontact 3D optical imaging of blood flow distributions in animals and humans.* IEEE International Symposium on Signal Processing and Information Technology, 2018: p. 441–446.
9. Zhao, M., et al., *Noncontact speckle contrast diffuse correlation tomography of blood flow distributions in burn wounds: A preliminary study.* Mil Med, 2020. **185**(Suppl 1): p. 82–87.
10. Abu Jawdeh, E.G., et al., *Noncontact optical imaging of brain hemodynamics in preterm infants: A preliminary study.* Phys Med Biol, 2020. **65**(24): p. 245009.
11. Huang, C., et al., *Noninvasive noncontact speckle contrast diffuse correlation tomography of cerebral blood flow in rats.* Neuroimage, 2019. **198**: p. 160–169.
12. Huang, C., et al., *Speckle contrast diffuse correlation tomography of cerebral blood flow in perinatal disease model of neonatal piglets.* J Biophotonics, 2021. **14**(4): p. e202000366.
13. Mazdeyasna, S., et al., *Intraoperative optical and fluorescence imaging of blood flow distributions in mastectomy skin flaps for identifying ischemic tissues.* Plast Reconstr Surg, 2022.
14. Huang, C., et al., *Alignment of sources and detectors on breast surface for noncontact diffuse correlation tomography of breast tumors.* Appl Opt, 2015. **54**(29): p. 8808–8816.

15. Ziegelitz, D., et al., *Pre-and postoperative cerebral blood flow changes in patients with idiopathic normal pressure hydrocephalus measured by computed tomography (CT)-perfusion.* J Cereb Blood Flow Metab, 2016. **36**(10): p. 1755–1766.

16. Volpe, J.J., et al., *Positron emission tomography in the newborn: Extensive impairment of regional cerebral blood flow with intraventricular hemorrhage and hemorrhagic intracerebral involvement.* Pediatrics, 1983. **72**(5): p. 589–601.

17. Ment, L.R., et al., *Alterations in cerebral blood flow in preterm infants with intraventricular hemorrhage.* Pediatrics, 1981. **68**(6): p. 763–769.

18. Proisy, M., et al., *Changes in brain perfusion in successive arterial spin labeling MRI scans in neonates with hypoxic-ischemic encephalopathy.* Neuroimage Clin, 2019. **24**: p. 101939.

19. Giesinger, R.E., et al., *Hypoxic-ischemic encephalopathy and therapeutic hypothermia: The hemodynamic perspective.* J Pediatr, 2017. **180**: p. 22–30.

20. Tortora, D., et al., *Regional impairment of cortical and deep gray matter perfusion in preterm neonates with low-grade germinal matrix-intraventricular hemorrhage: an ASL study.* Neuroradiology, 2020. **62**(12): p. 1689–1699.

21. Aquilina, K., et al., *A neonatal piglet model of intraventricular hemorrhage and posthemorrhagic ventricular dilation.* J Neurosurg, 2007. **107**(2 Suppl): p. 126–136.

22. Tang, T. and R.J. Sadleir, *Quantification of intraventricular hemorrhage with electrical impedance tomography using a spherical model.* Physiol Meas, 2011. **32**(7): p. 811–821.

23. Tang, T., et al., *In vivo quantification of intraventricular hemorrhage in a neonatal piglet model using an EEG-layout based electrical impedance tomography array.* Physiol Meas, 2016. **37**(6): p. 751–764.

24. Barbier, A., et al., *New reference curves for head circumference at birth, by gestational age.* Pediatrics, 2013. **131**(4): p. e1158–e1167.

25. Amiel-Tison, C., J. Gosselin, and C. Infante-Rivard, *Head growth and cranial assessment at neurological examination in infancy.* Dev Med Child Neurol, 2002. **44**(9): p. 643–648.

26. Li, Z., et al., *A statistical skull geometry model for children 0–3 years old.* PLoS One, 2015. **10**(5): p. e0127322.

27. Bolander, R., et al., *Skull flexure as a contributing factor in the mechanism of injury in the rat when exposed to a shock wave.* Ann Biomed Eng, 2011. **39**(10): p. 2550–2559.

28. O'Reilly, M.A., A. Muller, and K. Hynynen, *Ultrasound insertion loss of rat parietal bone appears to be proportional to animal mass at submegahertz frequencies.* Ultrasound Med Biol, 2011. **37**(11): p. 1930–1937.

29. Bonaroti, A., et al., *The role of intraoperative laser speckle imaging in reducing postoperative complications in breast reconstruction.* Plast Reconstr Surg, 2019. **144**(5): p. 933e–934e.

30. To, C., et al., *Intraoperative tissue perfusion measurement by laser speckle imaging: A potential aid for reducing postoperative complications in free flap breast reconstruction.* Plast Reconstr Surg, 2019. **143**(2): p. 287e–292e.

31. Frey, J.D., et al., *Mastectomy flap thickness and complications in nipple-sparing mastectomy: Objective evaluation using magnetic resonance imaging.* Plast Reconstr Surg Glob Open, 2017. **5**(8): p. e1439.

Summary and Future Perspectives

4

Within this book, it is hopeful that the reader has obtained sufficient information to both understand the role of speckle contrast diffuse correlation tomography (scDCT) among other imaging systems and gained perspective on its application potential. The evolution of scDCT instrumentation was outlined starting from its early partial contact instantiation as a PoC and leveraging its ncDCT heritage. Coverage continued with theoretical descriptions on the nascent flow acquisition method combining aspects from wide-field illumination/camera systems and point-like sources/APDs. The basics of fully non-contact scDCT instrumentation were itemized, related, and extended upon in continuously revised versions. Validating and characterizing expectations regarding these various scDCT incarnations was presented through analyzing computer simulation and tissue-like phantom measurements. The suitability of scDCT was expressed in terms of numerous specific applications which highlight upon the problems which may be solved. These include sites that are sensitive to contact such as mastectomy skin flaps and burn/wound tissues. They also include regions difficult to probe noninvasively such as the brain due to the skull and depth. Subject sizes ranged greatly, from humans down to mice and ages from adults to preterm infants. Sites tested included forearms, cortex tissues, deep wounds, and surgical breast reconstruction areas. These were all carried out with a noninvasive, portable, and relatively low-cost BFI device with the features to support battlefield, NICU bedside, animal facilities, and other constrained environments.

There are several limitations with scDCT and related future (or completed) directions. Some limitations from early versions were addressed and already set forth herein. Light will be shed on some of the most important of those remaining. As pointed out in discussions of scDCT operation, and touched upon within the applications themselves, the scan time of scDCT depends on several factors. There exist scenarios where the tradeoff between acquiring a dense set of boundary data encumbers the time required to do so such that rapid changes are missed or otherwise reduced. In these situations, the current scDCT iterations may not be capable of detecting the physiological changes of interest. To remedy this issue and find fast-changing signals, we recently designed a moving window reconstruction algorithm that increased the effective sampling rate of images. Another concern is that the optical properties are known to have an influence on flow calculations [1]. These are often assumed from literatures. Recently, a two-step fitting algorithm successfully extracted tissue absorption and blood flow (αD_B blood flow index) simultaneously using the scDCT arrangement [2]. This solution can assist in improving flow reconstruction accuracy. Additionally, with multiple source wavelengths, tissue oxygenation and derivatives (e.g., tissue oxidative metabolism) could be measured. The long reconstruction time for generating 3D rCBF distributions is reiterated here. A 2D mapping was noted to overcome this issue as feedback on the status of ongoing experimentations (or clinical measurements) is often desired if available. Another important benefit of this 2D mapping is that different S-D separations can be selected in order to achieve better depth sensitivity than with DCS or LSCI. Other proposed solutions to facilitate quick updates have involved task parallelization, GPU-based solvers, Jacobian subsets, and supercomputers [3,4]. The final topic is on the portability and scDCT instrumentation size. Compared to many other contemporary imaging modalities scDCT excels in this area. However, despite that improvement, it is still missing important shorthanded arenas. Toward this end, a miniaturized, fiber-free, wearable system based on scDCT and NIRS termed diffuse speckle contrast flow-oximetry has been recently developed for continuous tissue hemodynamic monitoring in freely behaving subjects [5–7]. This system shows promise for the efforts portrayed in this book to one day soon be involved in research on small moving rodents, ambulating hospital patients, and many other exciting directions.

REFERENCES

1. Irwin, D., et al., *Influences of tissue absorption and scattering on diffuse correlation spectroscopy blood flow measurements.* Biomed Opt Express, 2011. **2**(7): p. 1969–1985.
2. Zhao, M., et al., *Extraction of tissue optical property and blood flow from speckle contrast diffuse correlation tomography (scDCT) measurements.* Biomed Opt Express, 2021. **12**(9): p. 5894–5908.
3. Doulgerakis, M., et al. *Towards real-time functional human brain imaging with diffuse optical tomography.* in Proceedings of SPIE-OSA. 2017. CA.

4. Wu, X., et al., *Fast and efficient image reconstruction for high density diffuse optical imaging of the human brain.* Biomed Opt Express, 2015. **6**(11): p. 4567–4584.
5. Huang, C., et al., *Low-cost compact diffuse speckle contrast flowmeter using small laser diode and bare charge-coupled-device.* J Biomed Opt, 2016. **21**(8): p. 80501.
6. Huang, C., et al., *A Wearable Fiberless Optical Sensor for Continuous Monitoring of Cerebral Blood Flow in Mice.* IEEE J Sel Top Quantum Electron, 2019. **25**(1): p. 1–8.
7. Liu, X., et al., *Simultaneous measurements of tissue blood flow and oxygenation using a wearable fiber-free optical sensor.* J Biomed Opt, 2021. **26**(1): p. 012705.

Index

Note: **Bold** page numbers refer to tables and *italic* page numbers refer to figures.

Printed in the United States
by Baker & Taylor Publisher Services